Routledge Introductions to Development

Series Editors:
John Bale and David Drakakis-Smith

Primary Resources and Energy in the Third World

WITHDRAWN

N 0016805 X

In the same series

John Cole
Development and Underdevelopment
A profile of the Third World

David Drakakis-Smith
The Third World City

Allan and Anne Findlay
Population and Development in the Third World

Avijit Gupta
Ecology and Development in the Third World

John Lea
Tourism and Development in the Third World

Forthcoming

Rajesh Chandra
Industrialization and Development in the Third World

Chris Dixon
Agricultural Systems and Rural Development in the Third World

Graeme Hugo
Population Movements and the Third World

Janet Momsen
Women and Development in the Third World

Peter Rimmer
Transport Patterns in the Third World

Steve Williams
Global Interdependence
Trade, Aid and Technology Transfer

E-90/2

John Soussan

Primary Resources and Energy in the Third World

ROUTLEDGE
London and New York

First published in 1988 by
Routledge
11 New Fetter Lane, London EC4P 4EE

Published in the USA by Routledge
in association with Routledge, Chapman and Hall, Inc.
29 West 35th Street, New York NY 10001

© 1988 John Soussan

Typeset by Hope Services
Printed in Great Britain
by Richard Clay Ltd, Bungay Suffolk

All rights reserved. No part of this book may be reprinted or reproduced or utilized in any form or by any electronic, mechanical or other means, now known or hereafter invented, including photocopying and recording, or in any information storage or retrieval system, without permission in writing from the publishers.

British Library Cataloguing in Publication Data

Soussan, John
Primary resources and energy in the Third World.——(Routledge introductions to development).
1. Developing countries. Natural resources. Exploitation
I. Title
333.7′09172′4

ISBN 0–415–00672–4

Library of Congress Cataloging-in-Publication Data

Soussan, John
Primary resources and energy in the third world / John Soussan.
p. cm.—(Routledge introductions to development)
"Published in the USA by Routledge in association with Routledge, Chapman and Hall"—p.
Bibliography: p.
Includes index.
1. Mines and mineral resources—Developing countries. 2. Power resources—Developing countries. I. Title. II. Series.
TN127.S65 1988
333.79′09172′4—dc19

ISBN 0–415–00672–4 (pbk.)

Contents

Acknowledgements

The author wishes to thank Chris Holland for typing the manuscript, Sheila Dance and Chris Howitt for drawing the figures and Erika Meller for the photographic work. In addition, he would like to thank Peter Odell and Phil O'Keefe for the information and observations they provided.

1 Development and resources

This book is about resources and development. It looks at these two concepts (both of which are broad, and each of which has many definitions) within the context of the contemporary Third World. This context is familiar; news of the problems facing these countries is with us daily and concern is widespread. The basic problem is poverty (and the common, and misleading, image is starvation – in the Third World few people are really starving, most are 'just' poor), but issues such as inequality, economic stagnation, environmental destruction and others go hand in hand with it. These problems breed conflicts which in turn lead to greater inequalities, stagnation and destruction and hence worsening poverty. The gap between the 'haves' and the 'have-nots' of the world's system widens all the time and, whilst the proportion (as a percentage of total population) of the Third World's population which lives in absolute poverty has fallen in recent decades (the development success story), their absolute number has increased (the continuing development tragedy).

Of course, not all is gloom and doom. Some Third World countries have achieved significant increases in the living standards of most of their populations. These include such diverse countries as proudly socialist Cuba and unashamedly capitalist South Korea. In some other countries success is more patchy but still notable – India, most of South-east Asia and several Latin American countries could be described in this way. For some of these countries, however, there has been a false dawn of rapid economic growth followed by economic stagnation or decline. Elsewhere the story is of widespread deterioration of living standards and of the resource base surrounding small islands of progress or prosperity.

Economic growth may occur, but development which is truly sustainable, and which both reaches the poor and is ecologically sound, is less common.

As we shall see in this book, many forms of economic growth are based on patterns of resource exploitation which create as many problems as they solve. The idea of 'sustainable development' is currently highly fashionable and is indeed a worthy goal to strive for. How it can be achieved, when so many forces operate against it, is a more challenging problem. A recent BBC television series and accompanying book (Timberlake 1987), 'Only One Earth', sought and found examples of sustainable development, but only with difficulty and under special circumstances which must limit the extent to which such models could be generalized. Despite this, the forceful arguments put forward by the Brundtland Commission (1987), in their report *Our Common Future*, lead to inescapable conclusions: current trends in the world economy are not sustainable and, in particular, the way we view both development and resources must change if the current bleak outlook is to improve and genuine, sustainable progress for all of the world's people occur.

The above comments form the context within which this book is placed. In the following sections we look at two of the many resource issues which confront the Third World today. These are mineral exploitation in section 1 and energy in section 2. Both are vitally important, and the trends in the development of each of them are widely considered to have adverse impacts upon the countries involved.

Part I, which looks at mineral resources (including fuel minerals such as oil) is mainly focused at an international level, exploring the ways in which mining activities integrate mineral producers into the world economy and examining who benefits from this industry and why. Part II is concerned primarily with the way in which energy affects the welfare of people at a local level and, in particular, describes the complex interrelationships of economy and environment and the place of energy within these complex relationships. The two sections consequently deal with very different scales of operation. This provides an opportunity to discuss both questions of resource exploitation and wider development theories at these different scales, and through this, one hopes, to illustrate where and how the problems facing the Third World originate.

Before discussing in greater detail the distribution of the worlds' mining industry, it will be useful to look at some definitions of the concepts we shall be considering.

The nature of resources

A resource is anything that contributes to the process of production. As we shall see, resources take many forms, but all reflect a relationship between human wants, abilities and appraisal of the physical universe. In other words, a resource is a resource only within the confines of the needs and objectives of a given economic system, and more specifically of the social, technological, economic and institutional characteristics of that system. In an economic sense, resources do not exist; they are created.

Resources can be classified in a number of ways. There is not room here for a full discussion of these issues (see Blunden 1985, Fernie and Pitkethly 1985), but a first useful distinction can be made between *flow resources* (resources, such as soils and forests, whose availability can be depleted, sustained or increased by human actions), *stock resources* (non-renewable except over geological time – basically minerals) and *continuous resources* (which are available on a continuous basis and independent of man's action – resources such as solar and tidal power). This three-fold categorization is based on the physical properties of the resources. A second way of classifying resources is on their basic economic characteristics. Here it is crucial to distinguish between resources which are *commodities* (that is, subject to private property rights or possessing 'exchange value') and those which are *free goods* (non-commercial, or having no exchange value and not owned by anyone).

There is a degree of overlap between these two ways of classifying resources. Most continuous resources (such as sunlight) are free goods, whilst most stock resources and many flow resources are commodities in the modern world. However, this varies between places and over time. For example, forests in some places are subject to property rights and in others are not. In the past many more resources were free goods, but the extension of the global capitalist economy has led to the 'commodification' of many previously free resources. In other words, resources which previously had no exchange value have become commercialized as the world's economic system has developed and changed. Their commodification is a direct consequence of changes in the economic system of the place in which they are found. As an area becomes absorbed into the international economic system, the ways in which resources are used, owned and controlled changes as the local economic system changes. What was free now has value. This change in the nature of control of resources reflects the 'transition in modes of production' occurring throughout the modern Third World.

In this book the first part deals exclusively with minerals (stock resources), and in particular with the international markets for these commodities, the structure of the mining industry and the impact of this industry on Third World countries.

Part I
Minerals and development

2
Minerals: basic issues

Resources, reserves and recovery rates

Mineral resources are, as we have seen, stock resources; their supply is finite and in principle the global stock of a mineral diminishes as it is used. These stocks occur in finite quantities in fixed places – gold, as they say, is where you find it. So is oil, copper, cobalt, coal . . . This has a number of implications:

1 Mineral deposits exploited now will not be available in the future. This introduces the concept of *scarcity* associated with resource depletion.
2 There are unavoidable costs associated with finding, recovering, processing and transporting minerals. These vary according to the quality, location and geological characteristics of the deposit, as well as the labour costs and the technology available for its exploitation.
3 The price of minerals on the world market reflect both the purpose for which they are needed (their use value) and perceptions of the quantity available both now and in the future (their scarcity or exchange value).

This picture makes predicting the future of minerals all sound very straightforward. Unfortunately the story is not that simple. In principle, there is a finite amount of a mineral resource available in the earth's crust, but in practice the extent of that quantity is a matter of conjecture. This is partly because most of the earth's crust has not yet been fully explored, measured and inventoried, partly because minerals are not found in nice neat, homogeneous packages, but are rather mixed in with all sorts of other materials and partly because economic and technical constraints limit the proportion of a mineral deposit which can be exploited (the 'recovery rate'). All of this leads to massive uncertainty over the quantity of the various mineral resources which is likely to be available – they are finite, but how finite?

There are three standard definitions of the quantity of a mineral. These definitions measure very different things and are frequently confused. This is particularly true of many of the more pessimistic predictions of the world's future (the notorious Club of Rome report, *The Limits to Growth*, is a classic example of this). The definitions are as follows:

Proven reserves

These are deposits of a mineral which have been explored, measured and inventoried, and which are both recoverable with current technology and economically viable at current prices. In other words, proven reserves are deposits which we know are there and we know we can get at. For all minerals, proven reserves constitute only a small fraction of the material in the earth's crust.

Resource base

This is the opposite end of the spectrum: an estimate of the total quantity of a given material contained in the earth's crust. It is a highly theoretical concept which, for many minerals, provides no indication of the quantity which is likely ever to be available to man. For example, on average a cubic kilometre of rock contains 239 million metric tons of aluminium, 149 million metric tons of iron, 62 million metric tons of magnesium, nearly 3 million metric tons of manganese, 239,000 metric tons of zinc, 155,000 metric tons of copper and smaller quantities of many other minerals. The problem is that most of these minerals are normally present at concentrations which are less than a thousandth of those necessary to make their extraction viable. It is only where higher concentrations are found that mining is feasible and the deposits can be considered an ore. Nevertheless the notion of the resource base for minerals does give some indication of the abundance of many supposedly scarce minerals.

Ultimately recoverable resource

This concept is the most useful of the three as an indicator of the quantity of minerals which will be available to man. The meaning is clear: the amount of a mineral which will be recovered, given certain assumptions about future extraction and processing technologies and future costs and prices. The concept originated from the US Geological Survey, who defined it as 0.01 per cent of the material in the resource base. Since then the concept has been refined. The key issue in measuring the ultimately recoverable resource is the set of assumptions about future discovery rates, technological developments, commodity prices, etc. on which the estimate is based. As such the concept is highly elastic and technical. Given this, it is hardly surprising that forecasters and policy-makers tend to prefer proven reserves – they dislike the uncertainty and prevarication which surround ultimately recoverable resource measures. 'Proven reserves' is useless as a measure of future resource availability, however, as it simply presents a snapshot of the current position. A future tense must take account of technical and economic changes and future discoveries. These relationships are displayed diagramatically in figure 2.1.

TOTAL RESOURCES

IDENTIFIED
UNDISCOVERED
Demonstrated
Measured
Indicated
Inferred
HYPOTHETICAL (In known districts)
SPECULATIVE (In undiscovered districts)
ECONOMIC
SUBECONOMIC
Paramarginal
Submarginal
RESERVES
RESOURCES
Increasing degree of economic feasibility

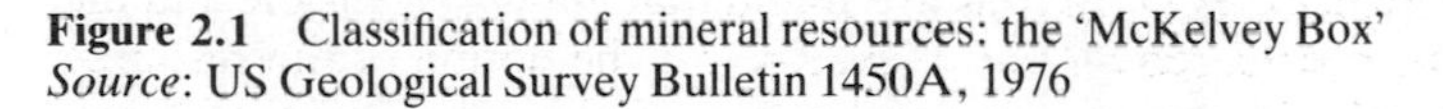

Figure 2.1 Classification of mineral resources: the 'McKelvey Box'
Source: US Geological Survey Bulletin 1450A, 1976

Tanzer (1980) gives an example of the differences between proven reserves and resource base (see table 2.1). These differences are startling. For example, at the 1974 rate of extraction the *proven reserves* of copper would only last 56 years and cobalt 97 years. In contrast, the *resource base* for these two minerals would last 216 million years for copper and 23,800 million years for cobalt. Of course, both sets of figures are meaningless. For all of these minerals, the ultimately recoverable resource is considerably greater than the proven reserves, but only a fraction of the resource base.

For most minerals, the size of the proven reserves changes constantly according to discovery and depletion rates. When prices and demand are high, more exploration occurs and reserves grow. Conversely, when prices are low (or uncertain) and supplies abundant, exploration and development efforts retract and the rate of addition to reserves declines. Odell (1986)

Table 2.1 Resource base and proven reserves for selected minerals

Life expectancies of world reserves for selected mineral commodities

Mineral commodity[a]	*1974 reserves (metric tons)*	*Average annual production 1972–4*	*Life expectancy in years at four growth rates*				*Average annual production growth 1947–74 (%)*
			0%	*2%*	*5%*	*10%*	
Bauxite (ore)	15.7×10^9	69.7×10^6	226	86	51	33	9.8
Chromium (ore)	1.7×10^9	6.5×10^6	263	93	54	35	5.3
Cobalt (Co)	2.4×10^6	25.3×10^3	97	54	36	25	5.8
Copper (Cu)	390.0×10^6	7.0×10^6	56	38	27	20	4.8
Gold (Au)	4.0×10^4	1.3×10^3	30	24	19	15	2.4
Iron (Fe)	87.7×10^9	0.5×10^9	167	74	46	30	7.0
Lead (Pb)	145.1×10^6	3.5×10^6	42	31	23	17	3.8
Mercury (Hg)	182.3×10^3	9.4×10^3	19	17	14	11	2.0
Nickel (Ni)	44.4×10^6	0.7×10^6	67	43	30	22	6.9
Tin (Sn)	9.9×10^6	0.2×10^6	42	31	23	17	2.7
Zinc (Zn)	118.8×10^6	5.6×10^6	21	18	15	12	4.7

[a] The notation in parentheses following the name of a mineral commodity indicates what the reserve and production figures actually measure. For example, for copper they measure contained metal (Cu), and for bauxite and chromium they measure ore (ore).

Life expectancies of resource bases for selected mineral commodities

Mineral commodity	*Resource base*[a] *(metric tons)*	*Average annual production 1972–4 (metric tons)*	*Life expectancy in years at four growth rates*				*Average annual production growth 1947–74 (%)*
			0%	*2%*	*5%*	*10%*	
Aluminium	2.0×10^{18}	12.0×10^{6}	166.0×10^{9}	1,107	468	247	9.8
Chromium	2.6×10^{15}	2.1×10^{6}	1.3×10^{9}	861	368	196	5.3
Cobalt	600.0×10^{12}	25.3×10^{3}	23.8×10^{9}	1,009	428	227	5.8
Copper	1.5×10^{15}	7.0×10^{6}	216.0×10^{6}	772	332	177	4.8
Gold	84.0×10^{9}	1.3×10^{3}	62.8×10^{6}	709	307	164	2.4
Iron	1.4×10^{18}	0.5×10^{9}	2.6×10^{9}	898	383	203	7.0
Lead	290.0×10^{12}	3.5×10^{6}	83.5×10^{6}	724	313	167	3.8
Mercury	2.1×10^{12}	9.4×10^{3}	223.5×10^{6}	773	333	178	2.0
Nickel	2.1×10^{12}	0.7×10^{6}	3.2×10^{6}	559	246	133	6.9
Tin	40.8×10^{12}	0.2×10^{6}	172.2×10^{6}	760	327	175	2.7
Zinc	2.2×10^{15}	5.6×10^{3}	398.6×10^{9}	1,151	486	256	4.7

[a] The resource base for a mineral commodity is calculated by multiplying its elemental abundance measured in grams per metric tons times total weight (24×10^{18}) in metric tons of the earth's crust. It reflects the quantity of that material found in the earth's crust.

Source: Tanzer (1980), 34–7

Table 2.2 World prices for selected minerals 1962–83

		Commodity prices (per metric ton: current price in US$, constant price in 1980 US$)							
		1962	*1965*	*1968*	*1971*	*1974*	*1977*	*1980*	*1983*
Copper	Current	644	1,290	1,241	1,080	2,059	1,309	2,183	1,592
	Constant	2,350	4,542	4,354	3,140	3,650	1,854	2,183	1,744
Lead	Current	154	317	240	254	593	618	906	425
	Constant	562	1,116	842	738	1,051	857	906	466
Tin	Current	2,471	3,893	3,126	3,501	8,201	10,762	16,775	12.988
	Constant	9,018	13,708	10,968	10,177	14,541	15,244	16,775	14,226
Zinc	Current	185	311	262	309	1,239	591	761	764
	Constant	675	1,095	919	898	2,197	837	761	837
Iron-ore	Current	10.8	10.1	8.4	10.5	12.8	13.4	17.6	21.6
	Constant	39.4	35.6	29.5	30.5	22.7	19.0	17.6	23.7

Source: World Resources Institute 1986, 232

shows the way in which oil-exploration efforts in the 1970s and early 1980s (after the oil-price rises) led to significant increases in the proven reserves of oil in the 1970–85 period despite the consumption in this period (see figure 2.2). Indeed, for every two barrels of oil used globally three were added to proven reserves. The dramatic drop in oil prices in early 1986 led to major retraction of exploration efforts, an event which will be reflected in declining reserves and, no doubt, fresh hysteria about oil running out in the last years of the 1980s.

The global picture for minerals is consequently one of uncertainty. For most minerals, the medium-to-long-term future is one of continued availability, but at a level of cost and in locations which are likely to be increasingly uncertain. Scarcity of minerals does exist, but this scarcity (both now and in the future) reflects economic (and, to an extent, technical) constraints, *not* the absolute constraint of the exhaustion of finite resources. The notion of scarcity is an important one, as the perception of imminent resource depletion and control over information about resource availability are vital aspects of the way in which the world's mineral markets are controlled.

The geography of mineral production

The geographical patterns of production inevitably vary from mineral to mineral, but for a large number of minerals some generalizations can be advanced which have widespread validity.

Firstly, the Second World (the Eastern bloc) is largely self-contained in terms of mineral production and demand. This is less true in the 1980s than it was in the past, but even today the Second World is not a major participant in the international minerals markets. The Soviet Union is a major mineral producer and contains a significant proportion of the world's reserves for a

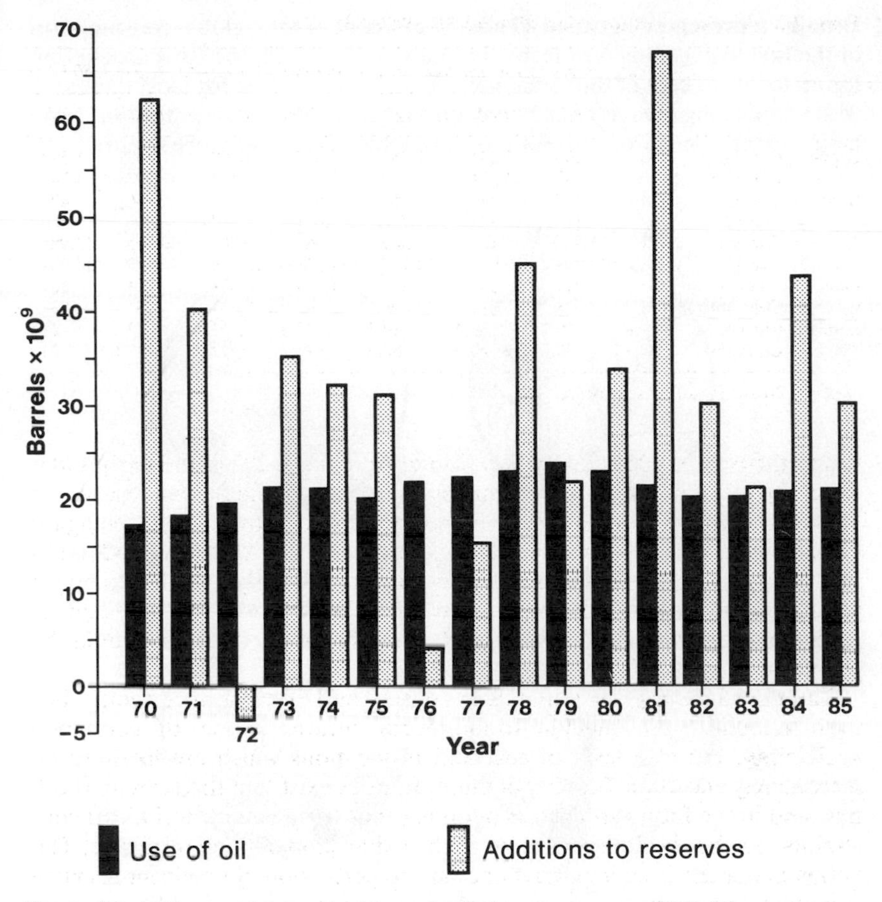

Figure 2.2 Oil: annual consumption and additions to reserves, 1970–85
Source: Odell 1986

number of minerals, a factor which will be of increasing importance in the future. The main concern of this section is the impact of the international minerals industry on the contemporary Third World, however, and as such, Second World mineral industries are only marginally of interest. The remainder of this section is consequently mainly concerned with relations between the First World and Third World.

Within this context, the bulk of demand for most minerals comes from a small number of advanced capitalist industrial nations, and in particular the United States, Japan and Western Europe. As figure 2.3 shows, six countries (the United States, Japan, West Germany, France, Italy and

Britain) represented between 49 and 57 per cent of total global consumption of the first five leading non-ferrous metals in 1983, with the USSR accounting for up to 20 per cent of the remainder. This pattern is true for most minerals, with some being characterized by even higher levels of concentration. (The major exception is iron, consumption of which is spread more evenly, with

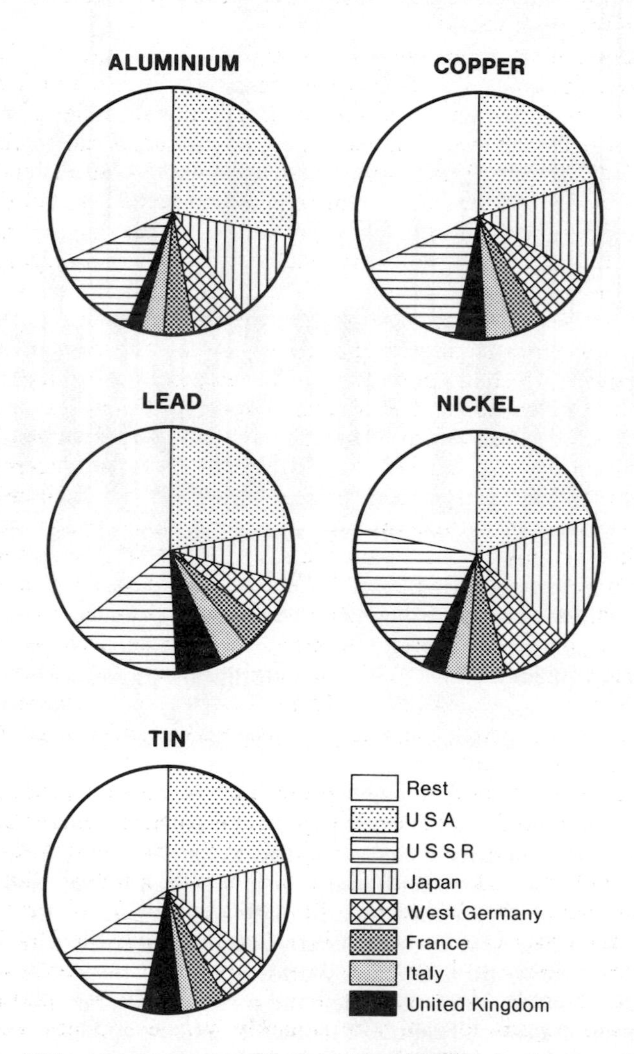

Figure 2.3 Major consumers of non-ferrous metals, 1983
Source: World Resources Institute 1986

Third World countries such as Brazil, India, Korea and others being major iron and steel producers.) This pattern of the concentation of demand for minerals reflects the structure of the world economy, with industrial production and consumer demand concentrated in the same group of affluent countries. As we shall see, it also reflects the investment strategies of the major mineral companies, which have tended to concentrate refining processes in the First World.

The production of many minerals is similarly concentrated, but in a different group of countries. This is the reason why international mineral markets are so important. As far as international mineral trade is concerned, two distinct groups of major mineral producers can be identified (see figure 2.4). The first is the United States, Canada, Australia and South Africa, all of which are capitalist countries which are either part of the First World or are closely allied to it. They all produce a wide range of minerals and have a set of interests which parallel those of the major consuming countries. The importance of these four countries in terms of mining investment has increased since 1970, with 80 per cent of mineral-exploration expenditure concentrated in them in the early 1970s. These countries have operated in a way which has prevented the development of producer cartels similar to the Organization of Petroleum Exporting Countries (OPEC) for other minerals. We shall return to these points in the next chapter.

The second group of countries is Third World mineral producers. There are many of these, but most depend on one, or at most two, main minerals, earnings from which form the basis of their export earnings and are the main source of income for the development strategies their governments are pursuing. For example, Indonesia, Bolivia and Malaysia are major tin producers, whilst Zambia, Chile, Zaire and Peru produce most of the copper mined in Third World countries. A country such as Zambia illustrates the problems Third World mineral exporters face. Over 90 per cent of her export earnings come from copper, a level of dependence which places her at the mercy of the international market and exaggerates the impact of price movements.

Within the Third World, a large proportion of mineral production is concentrated in a number of specific regions. Of particular importance are the Andean region and west coast of South America, Central Africa (in particular a belt through eastern Zaire and central Zambia), South-east Asia and, of course, the Gulf region of the Middle East for oil production. This does not mean, of course, that mineral production in the Third World is confined exclusively to these areas. There are major producers in other regions (for example, Jamaica for bauxite, Morocco for phosphates, Venezuela and Nigeria for oil) who similarly have economies which are largely dependent upon mineral export earnings.

There are also a number of other Third World countries which produce

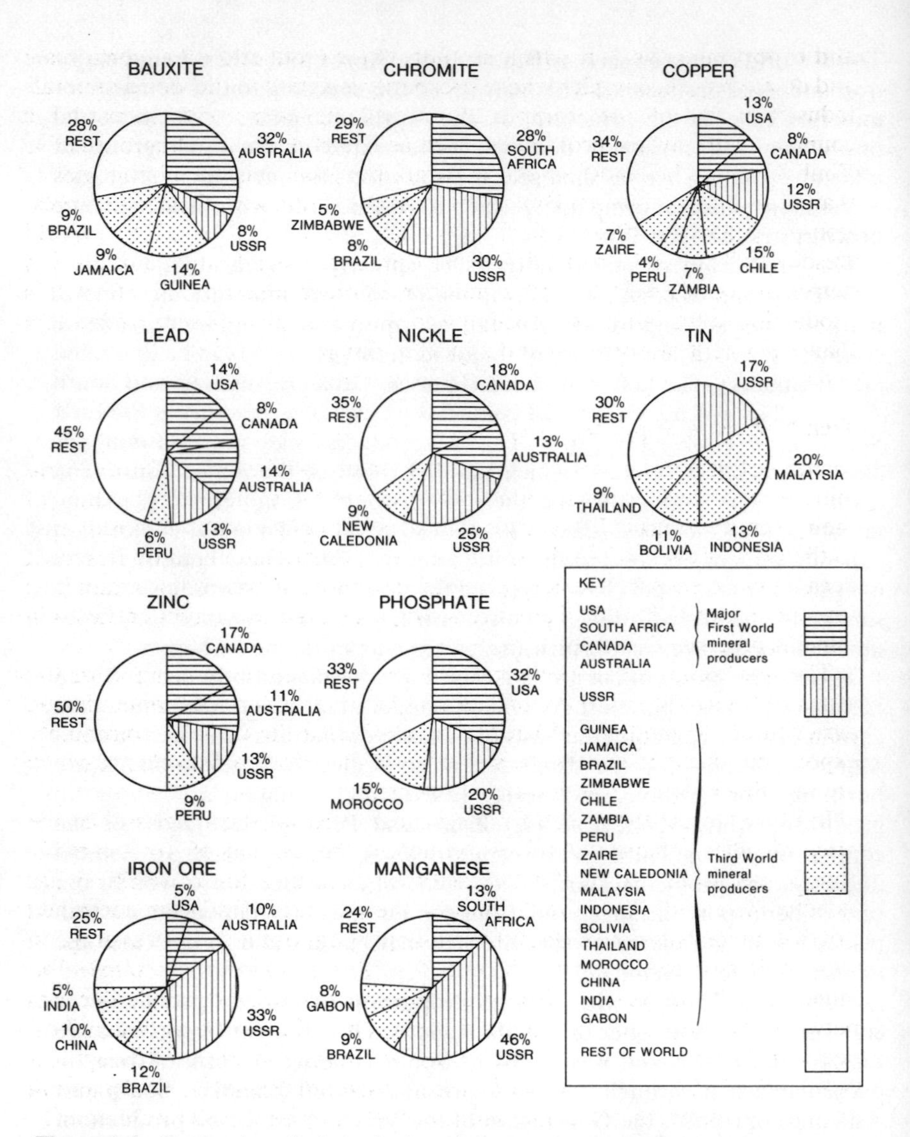

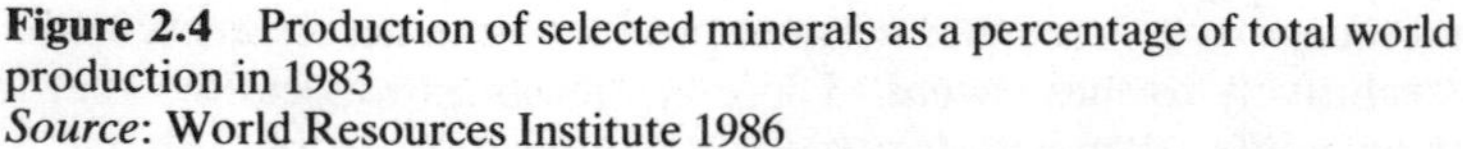
Figure 2.4 Production of selected minerals as a percentage of total world production in 1983
Source: World Resources Institute 1986

and export minerals, but which are not major producers on a world scale and do not have economies whose prosperity is as tied to the world minerals industry as that of, for example, Zaire or Chile. Some of these are large countries with diverse economies, such as Brazil or India. Others, such as Zimbabwe or Thailand, are not as large, but have smaller mining sectors than the major producers. Brazil, for example, is the world's second largest iron-ore producer (after the USSR), producing 89,000 tonnes in 1983 (World Resources Institute 1986). Brazil also produces significant quantities of petroleum, coal, bauxite and a number of other minerals. Much of this production is for domestic consumption, however, and minerals form less than 1 per cent of both GNP and export earnings.

Trends in mineral prices

In the last section it was explained that many of the main Third World mineral exporters depend upon these exports for a major proportion of their export earnings. This means that their economic health is closely tied to the price of their mineral exports in the international markets. In recent decades two trends have emerged in the international price of most minerals (except oil) which have had an adverse impact on the economies of the main mineral producers.

The first trend is that there has been a long-term deterioration of the terms of trade of commodity producers. This is true for both minerals and agricultural commodities. What this means is that the value of commodity exports has declined relative to the value of manufactured products, which are mainly exported from industrial countries.

In other words, the price of minerals has risen far less quickly than the price of other products. As a consequence, the amount of, for example, copper ore needed to pay for a tractor, a truck or a ton of fertilizer has steadily increased. As figure 2.5 shows, the general trend in the combined price of minerals is downwards (from 149 per cent of the 1977–9 average in 1964 to 91 per cent in 1984). The details of this trend vary from mineral to mineral, with some such as iron ore and copper showing dramatic declines between the mid 1960s and the mid 1980s, and others such as zinc and tin declining less rapidly or even marginally increasing in value (although, as case study A shows, the price of tin collapsed after 1984). This decline in prices has spectacularly continued since 1984. The *Financial Times* (14 January 1987) showed that non-fuel commodity prices declined by 25 per cent between mid 1984 and the end of 1986, reaching their lowest level in real terms since the worst days of the Depression of the 1930s.

The second trend in international mineral prices has been that in the short term they have been highly volatile. Again the trend has varied slightly for different minerals, but violent rises and falls in price over the short term

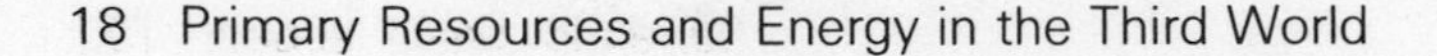

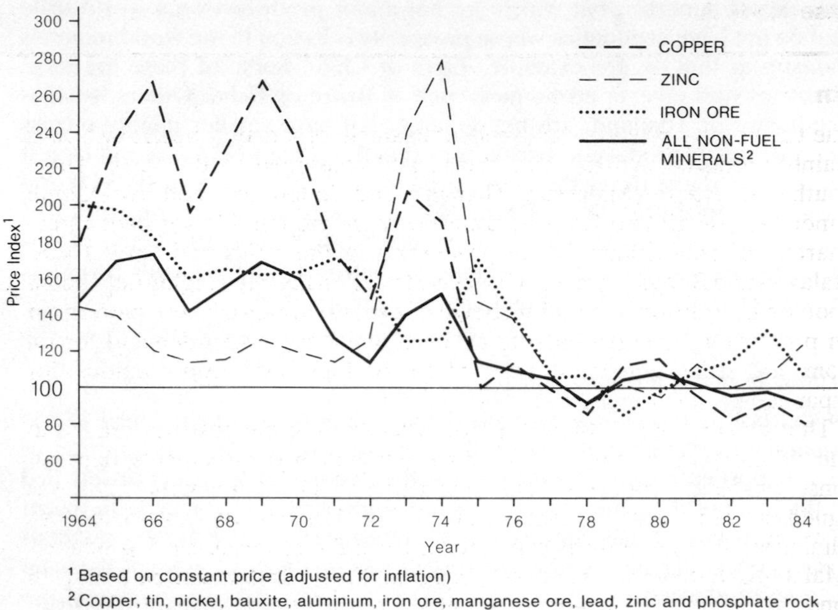

¹ Based on constant price (adjusted for inflation)

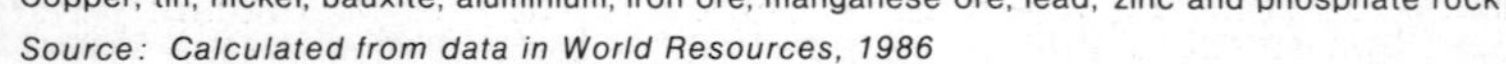

² Copper, tin, nickel, bauxite, aluminium, iron ore, manganese ore, lead, zinc and phosphate rock

Source: Calculated from data in World Resources, 1986

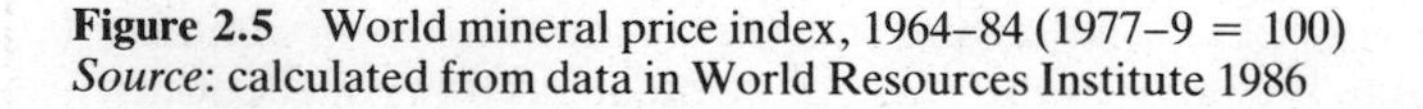

Figure 2.5 World mineral price index, 1964–84 (1977–9 = 100)
Source: calculated from data in World Resources Institute 1986

have characterized virtually all of them. The figures in table 2.2 and figure 2.5 show this on a year-by-year basis, with the general downward movement over time distorted by spectacular peaks and troughs for both individual minerals and the cumulative index of major non-fuel minerals. The data presented in table 2.2 and figure 2.5 are annual averages. Price fluctuations are even more volatile in the short term. For example, the collapse of the International Tin Council in late 1985 resulted in tin prices falling from US$12,000 per tonne in October 1985 to just over US$5,000 per tonne in May 1986 (see case study A).

Case study A

Tin

The tin industry provides an excellent example of a major mineral which is mainly produced in the Third World and consumed in the First. Three South-east Asian (Malaysia, Thailand and Indonesia) and two South American (Brazil and Bolivia) countries were responsible for over three-quarters of non-communist tin production in the 1970s and early 1980s. Malaysia and Bolivia, in particular, have traditionally been highly dependent upon tin production as a major source of export earnings. These patterns of tin production emerged initially in the colonial era, when demand for tin from industrial countries was extremely high and opportunities for expanded production great.

The changing nature of North–South relations in the post-war era led to one of the few examples of successful commodity agreements between consumers and producers. The International Tin Agreement (ITA) was signed in 1956 and the International Tin Council was established to administer it. The council contained 22 members, including 6 producers (Malaysia, Indonesia, Thailand, Nigeria, Zaire and Australia) and 16 consumers, including Japan and members of the EEC. Significantly Bolivia, Brazil and the United States were never members.

The ITA operated more or less successfully for nearly 30 years, using buffer stocks and export controls to keep international tin prices between a floor price of US$11.7 per kg and a ceiling price of US$15.2 per kg. The ITA succeeded in avoiding some of the worst fluctuations in commodity prices until the early 1980s, when pressures upon it led to its collapse and a rapid downward spiral in tin prices. It worked for so long because it protected both producers (who were guarded from rapidly falling prices) and consumers (who were protected from high prices or inadequate supplies). This was particularly true through the 1970s, when most mineral prices were undergoing the most violent fluctuations. In this period the failure of non-fuel-producer cartels to emulate OPEC's success contrasted with the effectiveness of the Tin Council in serving the interests of both producers and consumers.

By 1980, however, the ITA was beginning to look vulnerable, and during the 1980s pressure on tin prices grew further. There were two reasons for this. Firstly, non-member countries of the ITA expanded production rapidly (Bolivia, Brazil and China were particularly important in this) to a point in 1985 where they accounted for 40 per cent of world tin production. Secondly, demand for tin fell rapidly (from 180,000 tons in 1980 to 154,000 tons in 1985). This was partly because of the emergence of substitutes, such

Case study A (*continued*)

as aluminium and plastic, and partly because sustained economic recession across the First World led to a general decline in mineral demand.

The Tin Council continued to intervene by purchasing buffer stocks. The cost of maintaining these became increasingly expensive. By 1985 a stockpile of 65,000 tons (42 per cent of annual production) had accumulated. The crash, when it came, was spectacular. The critical point was reached when the Tin Council ran out of money to pay for its intervention operations. In late October 1985 tin trading was suspended on the London Metal Exchange (the main international metallic-minerals trading centre), which was thrown into the deepest crisis in its 107-year-old history when the Tin Council defaulted on £900 million owed to banks and metal traders. When tin trading recommenced in the spring of 1986 prices fell rapidly to the lowest levels since the 1930s. In May 1986 they were £3,500 per metric ton, a drop from the official pre-collapse price of £8,500 per metric ton. The impact on the main producers was traumatic. In Malaysia over 350 mines have been closed, 30,000 people thrown out of work and foreign exchange earnings severely affected. Bolivia's economy, already in crisis, nearly collapsed and is becoming increasingly dependent upon the illegal cocaine trade. Others suffered similarly. In October 1986 the ITA fell apart amidst a series of legal wrangles which will take years to sort out and which have severely dented confidence in hopes for commodity agreements for other minerals.

The reasons for these long- and short-term price fluctuations are complex. Briefly, they reflect cyclical fluctuations of growth and recession in the world economy in general and the major western industrial nations in particular; long-term changes in the nature of industrial production processes; and the effects of concern over and the failure of mineral-producer cartels (this last factor is particularly important in explaining the rapid rise and fall of commodity prices in the mid 1970s, following the first oil crisis). All of these issues are discussed at greater length in the next chapter.

The major exception to these price trends for mineral commodities is, of course, oil. This is because oil is so important as a component of the world's economy that it has a more complex relationship to economic fluctuations, and is the one commodity for which a producer cartel (OPEC) has had some measure of success. As a consequence, the real price of oil has increased over the last 20 years. These increases have been achieved by two rapid jumps – the oil crises of 1973/4 and 1979/80 (see figure 2.6). These increases

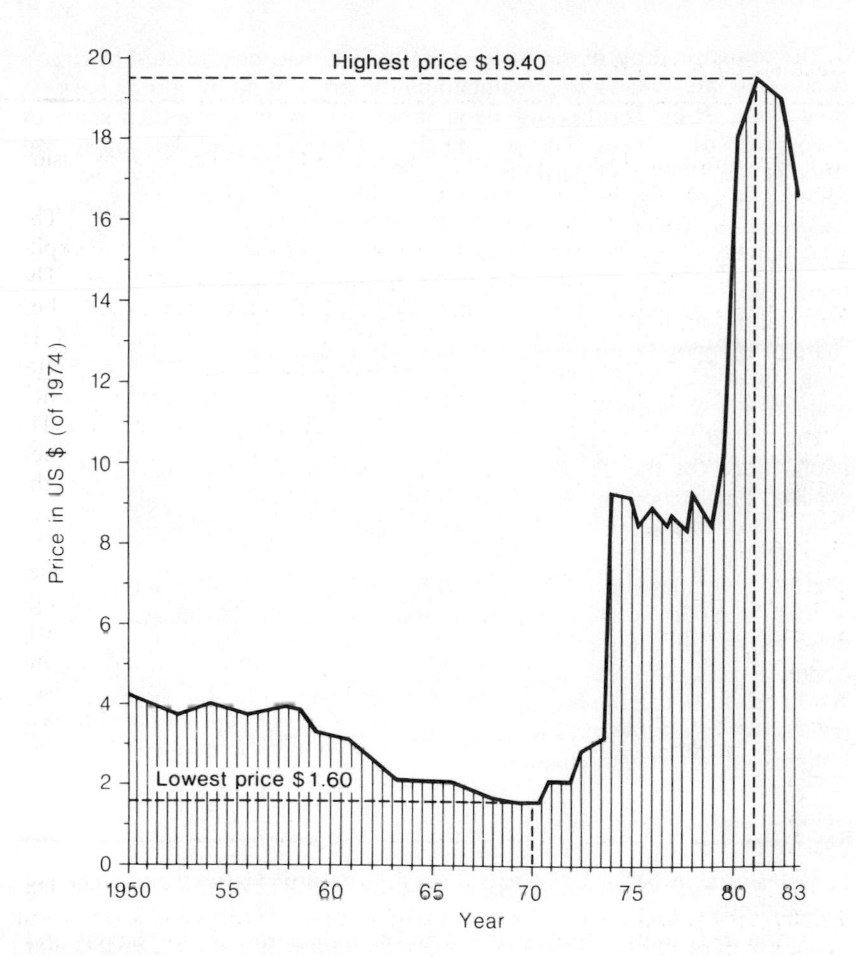

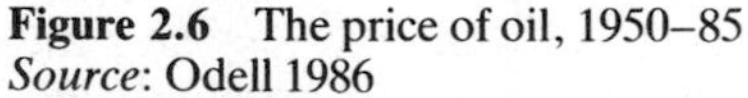
Figure 2.6 The price of oil, 1950–85
Source: Odell 1986

were achieved largely through the power of OPEC, the one (temporarily) successful example of a commodity-producer cartel. This success was relatively short-lived, however, as oil prices declined just as rapidly in the early months of 1986. The future of oil prices is uncertain, but Peter Odell (1986), one of the world's leading experts, suggests that a price between $10 and $20 per barrel is likely to remain for some time. This is considerably higher, in real terms, than the artificially low price of the 1960s, but much lower than the peak of the early 1980s. The decline in OPEC's power is probably permanent.

The dramatic drop in the price of oil in 1986 was precipitated by OPEC's decision in late 1985 to stop maintaining the price of oil by cutting back on production. It did this because its members had been losing their share of the world's oil market. This was partly due to depressed demand for oil through the 1980s and partly because of the emergence of major new oil production areas, such as the North Sea, Alaska and Mexico, in countries which are not members of OPEC. Exports of oil from OPEC members fell from 1,480 million tons in 1973 to 700 million tons in 1985. At the same time, production of non-OPEC oil outside Communist countries has increased from 760 million tons in 1973 to 1,225 million tons in 1985. This means that OPEC's share of the world's oil market (excluding the Second World) has declined from 66 per cent in 1973 to 36 per cent in 1985, a trend which has dramatically weakened its ability to control prices.

The trends in mineral prices discussed above have had profound implications for the Third World countries which depend upon mineral exports. For non-oil producers, the long-term decline in the value of their main source of foreign exchange earnings has had a disastrous impact upon their development prospects, and in particular on their ability to develop alternative sectors of their economy. The effect of this erosion of their terms of trade has been made worse by volatile short-term price fluctuations, as these make any attempt at economic planning impossible. As such, the trends in mineral prices in recent decades have worked against the Third World producers and in favour of the First World consumers. That this is so reflects the way the world's mineral industries are organized. This structure is discussed in the next chapter.

Key ideas

1 A resource is defined by societal need, economic viability and technical capability.
2 Resource scarcity reflects political and economic forces, not the physical stock of the resources.
3 The geography of the world's mineral industry reflects the geography of the world economy.
4 Mineral prices are volatile in the short term and are declining in the long term.

3 The international minerals industry and the Third World

Chapter 2 examined the basic characteristics of mineral resources and the geographical distribution of the world's minerals industry. The importance of the mining industry for the main Third World mineral producers was emphasized, and the adverse impact of trends in mineral prices discussed. In this chapter the reasons why mineral production is concentrated in certain countries and why the international minerals industry has operated against the interests of the Third World are discussed through a more detailed examination of the structure of the world's minerals industry.

Any discussion of international mining must centre on the three main groups of actors involved. These are:

1 Producers: the nations where mineral production is located. Chapter 2 indicated that production is concentrated in a limited number of countries. These can be sub-divided into First World producers (such as Canada) and Third World producers (such as Zambia).
2 Consumers: the nations who are the main source of demand for minerals. As we have seen, these are largely the industrial nations of the First World.
3 Production agents: the companies which actually mine, transport, process and sell the minerals. International production is dominated by a relatively small number of these. Most are giant multinational corporations (MNCS), for example, Exxon or Rio Tinto-Zinc, which have their headquarters in the First World. Of less but increasing importance are state-owned companies of the Third World producers, for example,

Zimco in Zambia. These state corporations are particularly important at the mining stage in the production process.

These three groups of actors have different sets of interests and varied forms of control. Broadly speaking, the *producers* are concerned with maximizing the price of the particular mineral they produce. This is particularly true for Third World producers, who also seek stability in prices and security in their markets. Such concerns are reflected in the development of state-owned production companies. The major First World consumers' attitude is more ambivalent, as they are also major consumers and, with the exception of South Africa, are far less dependent upon mineral production and exports.

The major *consumers*, western industrial nations, have opposite concerns. Their objectives are to secure supplies to meet existing levels of demand (which can change rapidly) at the lowest possible price. Where possible, industrial consumers seek to reduce demand or develop substitutes, and engage in strategies such as strategic stockpiling, agreeing long-term supply contracts and gaining direct control of supply sources in order to guarantee security of supplies. Although much of the demand in the First World comes from private industrial companies, the governments of these countries have long been concerned with mineral supplies and have been active in their involvement in the world's mineral industry.

The final group of actors, the *multinationals*, are the dominant force in the operation of the world's mining industry. Their main concern is to maximize long-term profits, an objective which includes policies designed to guarantee the security of investments and, wherever possible, to minimize the risks inherent in the highly competitive and unstable mineral markets in which they deal. These objectives are embedded in the corporate strategies mining MNCs pursue. The characteristics of these companies and of their strategies are of central importance to an understanding of the impact of mining on the Third World. We shall consequently look at them in more detail.

Before doing so, it will be useful to discuss briefly the uncertainties and costs which make mining such a risky business. These uncertainties are found at each stage of the production process. Some stem from the technical problems inherent in mineral production. Others reflect the structure of mineral markets and the political context within which mineral production occurs.

Mining: a risky business

The first stage in production, and first area of uncertainty, is the exploration for new deposits. This is highly technical and extremely expensive – the companies usually have to pay the host government for concessions (which carry with them the right to develop deposits discovered) and the exploration process itself costs a great deal. Only a small percentage of exploration efforts lead to the discovery of economically exploitable deposits. Technical developments (for instance, in the use of data from satellites) are reducing the uncertainties involved in exploration, but not the costs, and it still remains a risky business.

The second major area of uncertainty is over the long-term rate of return on mining investments. The size of investment necessary to develop a new mining complex is huge – in some cases billions of dollars, and rarely less than $100 million. The risks associated with such huge investments are compounded by the long development times involved. It is usually at least ten years between the discovery of a major mineral deposit and the first production. For example, the development of the Cuajore copper mine in southern Peru took 14 years to negotiate and develop, and involved an investment of $726 million.

Committing such huge sums for long periods before receiving any returns is both costly and risky. We have seen in chapter 1 that mineral prices are highly volatile in the short term and, for most minerals, have been declining in the long term. The dangers this entails for an MNC which is committing large sums for a number of years are obvious; declining prices may destroy the profitability of an investment before the first ton of ore is mined. MNCs consequently expect, and do everything they can to get, very high rates of return before they are willing to commit themselves to an investment.

The main reason for the volatile price movements of minerals is the exaggerated impact of fluctuations in the economies of First World countries on demand for mineral resources. Tanzer (1980) gives an example of this. A minor recession in the United States in 1970 led to a 0.3 per cent drop in GNP. This led to a 3.3 per cent decline in sales of consumer durables, such as refrigerators, demand for which is highly susceptible to economic fluctuations. This in turn led to a 12 per cent decline in demand for copper (one of the main markets for which is in consumer durables), which in turn led to a 34 per cent decline in the copper industry's overall profitability. What this means is that demand for minerals is closely tied to the upturns and downturns of the economies of the First World, and in particular of the United States. If a mine comes on-stream during a period of protracted recessions, such as the early 1980s, then its viability is threatened by the level of demand for the ore it produces. Mining companies are consequently concerned to control and protect the outlets for the minerals they produce.

The final major area of uncertainty stems from the politics of the world's mining industry. In particular, in recent decades many Third World mineral producers have attempted to change their relationship with the MNCs and to keep a greater proportion of the benefits from the exploitation of their resources for themselves. At its most extreme this has involved the wholesale nationalization of foreign-owned mining operations (action which is vigorously resisted by both the MNCs and First World governments) but also includes increased royalties and taxes, the formation of state-owned production or marketing organizations, co-operation between Third World producers of particular minerals and so on. These actions have led to increased uncertainty and greater concern over the safety of investments by MNCs in many regions of the Third World. These issues are looked at in greater detail later in this chapter.

The uncertainties listed above make mining, as we have seen, an extremely risky business. Such risks are compounded by the highly competitive nature of modern capitalism, and also by, in mining, the ever present danger that new deposits of much cheaper ores or new substitutes for minerals may be discovered. Many mining companies have failed, either to be swallowed up by a competitor or to disappear altogether. In many areas of mining these uncertainties are such that risk minimization has become the main concern of companies and profit maximization is a secondary consideration. This has led to a number of distinctive corporate strategies which have important implications for the relationship of MNCs to the Third World. These are examined next.

The mining multinationals

The major mining multinationals are economic giants. The largest oil producer, Exxon, is the world's biggest company and has an annual turnover of tens of billions of dollars, a level of economic activity which is bigger than the economies of all but a handful of the largest Third World countries (see table 3.1). The same is true for the other major oil producers

Table 3.1 Selected countries and mining corporations ranked by GDP and annual sales

Rank	*Company/Country*	*Sales (1983)/GDP (1984) (billion US$)*
1	Brazil	187.1
2	Mexico	171.3
3	India	162.3
4	Exxon (USA)	88.6
5	South Korea	83.2
6	Indonesia	80.6
7	Shell (Netherlands/UK)	80.6
8	Nigeria	73.5
9	Mobil (USA)	54.6
10	British Petroleum (UK)	49.2
11	Thailand	42.0

Rank	*Company/Country*	*Sales (1983)/GDP (1984) (billion US$)*
12	Texaco (USA)	40.1
13	Colombia	34.4
14	Malaysia	29.3
15	Standard Oil (USA)	27.6
16	Social (USA)	27.3
17	Gulf Oil (USA)	26.6
18	Atlantic Richfield (USA)	25.1
19	ENI (Italy)	25.0
20	Chile	19.8
21	Occidental (USA)	19.1
22	Peru	18.8
23	ELF (France)	18.2
24	Petrobras (Brazil)	16.3
25	Pemex (Mexico)	16.1
26	Tenneco (USA)	14.4
27	Morocco	13.3
28	Idemitsu Kosan (Japan)	10.8
29	Canadian Pacific (Canada)	10.4
30	Ecuador	9.9
31	Ashland Oil (USA)	7.9
32	Cameroon	7.8
33	Rio Tinto-Zinc (UK)	7.3
34	Imperial Oil (Canada)	7.2
35	Minnesota Mining (USA)	7.0
36	DSM (Netherlands)	6.9
37	Barlow Rand (S. Africa)	6.7
38	Nippon Mining (Japan)	5.6
39	Sri Lanka	5.4
40	Alcoa (USA)	5.3
41	Kenya	5.1
42	Continental (USA)	4.9
43	Zaire	4.7
44	Zimbabwe	4.6
45	Showa Oil (Japan)	3.9
46	Bolivia	3.6
47	Reynolds Metals (USA)	3.3
48	Kaiser Aluminium (USA)	2.9
49	Zambia	2.6
50	Noranda (Canada)	2.5
51	CSR (Australia)	2.5
52	Jamaica	2.4
53	Papua New Guinea	2.4
54	Amax (USA)	2.3
55	Imetall (France)	1.9
56	Phelps Dodge (USA)	1.0
57	Central African Republic	0.6

Note: This is *not* a comprehensive list.
Sources: Companies: 'Fortune 500', *Fortune Magazine* (1984)
Countries: World Bank (1986)

(the 'Seven Sisters', as Sampson calls them, all of which are based in the USA or Western Europe. Non-oil-mining companies are not in the same league as the oil giants, but are still extremely large. For example, Rio Tinto-Zinc had a turnover of about $8 billion in 1984, compared to Zambia's GNP of $2.6 billion and Zaire's of $4.7 billion (see case study B). The size of the mining MNCs is of central importance, as it gives them a degree of economic power which is central to their operation.

The growth in the size and power of MNCs has been a feature of the development of the world's economy since World War II. This is partly due to the growth in the international economy and partly due to the process of mergers and takeovers which now dominate business life. This is called a process of capital concentration. The growth in size of MNCs has made them increasingly complex, diverse and sophisticated organizations. This has been particularly crucial for mining MNCs, as it permits them to take the risks the mining industry entails. Mining is a risky business. The size of investments are huge and the uncertainties involved are great. The increasing size of MNCs has been one of the outcomes of these risks, as larger corporations are better able to survive in this risky and competitive business.

The size of the MNCs also puts them in a powerful position when it comes to negotiating exploration and mining concessions with Third World producer countries, as they have the resources and sophisticated organization which frequently permits them to dominate such negotiations. This was particularly true in the past, when mining MNCs virtually dictated terms and conditions which brought few benefits to the countries whose mineral deposits they were exploiting. The natural resentment of such deals was a powerful force behind the development of a much more aggressive stance by some producer countries, a stance which led in some cases to wholesale nationalization of mining production, the development of state-owned mining companies and the move towards producer cartels among Third World mineral-producing countries. We shall return to these issues below.

The growth of MNCs has been paralleled by their diversification. Large mining MNCs are increasingly involved in a range of economic activities, many of which lie outside the mining sector. This is known as *horizontal diversification* and frequently takes the form of companies moving into industrial processes related to the minerals they produce; for example, some oil companies make oil-based chemical goods such as plastics, paint, synthetic fibres and so on. It can also involve companies diversifying the minerals they produce, or even investing in industries unrelated to mineral production (see case study B).

A second strategy for decreasing risk, which growth has permitted many MNCs to adopt, is what is known as *vertical integration*. A mineral industry involves a series of distinct stages, each of which has costs, risks and profit opportunities associated with it. For example, the copper industry entails exploration for copper-ore deposits, development of a mine, the mining itself, initial processing of the ore, transportation of the concentrates, secondary smelting to produce raw copper and, finally, processing of the copper into products, such as copper wire, needed by the customers. Each of these stages may be located in different places. Vertical integration involves a mining company's expanding, for example, into transporting or refining the ore it extracts (or similarly a refining company starting to mine its own supplies). It is usually brought about by mergers or takeovers, rather than by a company starting a new operation from scratch.

These two corporate strategies, diversification of operations and vertical integration, are important features of the world's mining industries. They permit companies to cope with the risks inherent in mining and to generate far higher levels of profitability than would otherwise be the case. By diversifying, companies are able to spread risks and overcome price fluctuations, as their viability is not wholly dependent upon one product. Developing supply sources in a number of countries gives them greater leverage if any one supplier attempts to negotiate better terms. This is particularly true if the company has access to supplies in First World countries or Third World regimes whose governments are 'friendlier'.

Vertical integration has a number of advantages for MNCs. By 'selling' to themselves, they are guaranteed an outlet for the minerals produced (and a supply to their refineries) and have far greater control over the industry. In large-scale mining much of the cost is in the initial capital investment. Once a mine is developed, the marginal production or running costs for each ton of ore are low. Because of this, it is worthwhile for the company to carry on producing, whatever the price of the mineral, as most of their costs are fixed. By vertically integrating, mining MNCs can make sure that they have an outlet for their production.

Vertical integration also permits MNCs to take advantage of the profits to be made at each stage of the production process. Where different stages are located in different countries, they can also take advantage of different tax regimes by loading costs and minimizing profits where taxes are high, and inflating profits where taxes are low. In particular, it is normal to find production and processing in different places. Much of the processing of minerals mined in the Third World takes place in the First World in plants owned by MNCs. This is true even where the mining operation has been nationalized. There has been some growth of state-owned processing and refining plants in the Third World, but most minerals are still exported unprocessed or with only the initial stages of concentration done. A significant factor which limits the scope for Third World processing and mineral-based manufacturing is the export barriers First World countries (which are the main market) place upon the processed goods. There are no such barriers for most unprocessed minerals. Control of processing technology, shortages of capital for investment and the lack of local demand also restrict the scope for the development of processing capacity in developing countries.

Case study B

The Rio Tinto-Zinc Corporation

Rio Tinto-Zinc (RTZ) is one of the world's largest non-oil-producing mineral corporations. It is truly multinational, with operations which span all the world's continents except Antarctica (figures B.1 and B.2). It has a corporate structure which is characteristic of many MNCs, with a number of subsidiary companies operating in different spheres of activity (see table B.1). Mining and related activities form the heart of RTZ's operations, but RTZ companies are also involved in chemicals, engineering, construction and various servicing activities. It draws its profits from across the globe and was one of the first MNCs to pursue a policy of wide diversification of activities in order to spread its risks. This diversity and a vigorous management style permitted RTZ to come through the protracted recession of the early 1980s better than many comparable corporations.

In its annual reports and publicity material RTZ emphasizes strongly its image as a diverse, dynamic capitalist organization. As a private MNC, its aims are to maximize its profits under prevailing market conditions and to protect the interests of its shareholders. The company's literature also stresses what are presented as benign policies on environmental protection, workers' conditions and management and training practices. RTZ clearly recognizes the potentially devastating environmental impact of many of its operations around the world and claims to be doing all it can to monitor these impacts and minimize their effects. In its Third World operations, the company claims to provide wage levels and work conditions superior to those required by the laws of the countries in question, and to have comprehensive training programmes for local personnel at all levels. The picture the company presents is of an organization which operates vigorously but fairly within international market conditions to develop and process resources for the mutual benefit of the corporation, its employees and the different host countries in which they work.

Partizans is an umbrella organization of pressure groups, communities and shareholders opposed to RTZ's overseas activities. The story it tells contrasts sharply with the company's own literature. In its book, *RTZ Uncovered*, the organization claims that

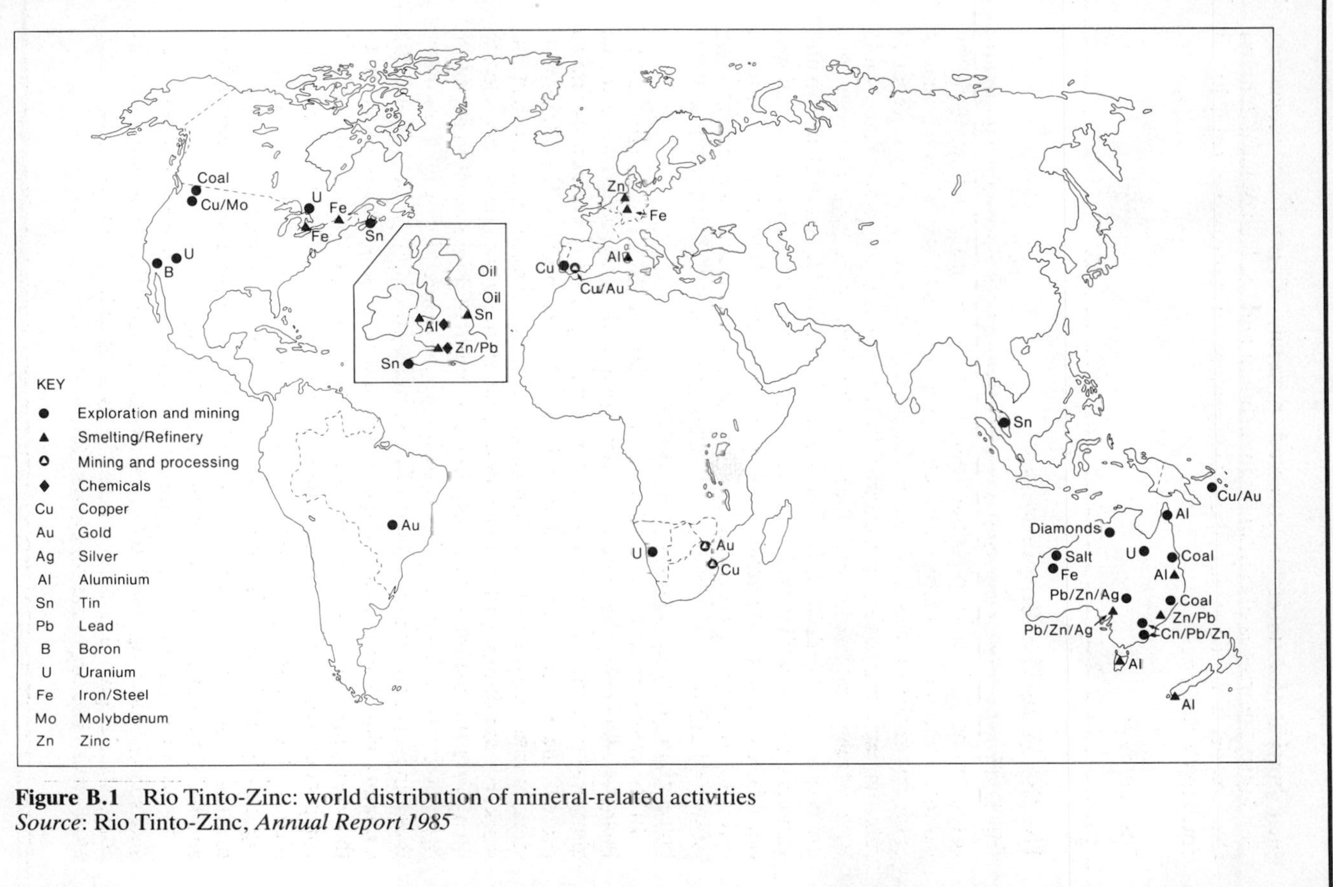

Figure B.1 Rio Tinto-Zinc: world distribution of mineral-related activities
Source: Rio Tinto-Zinc, *Annual Report 1985*

Case study B (*continued*)

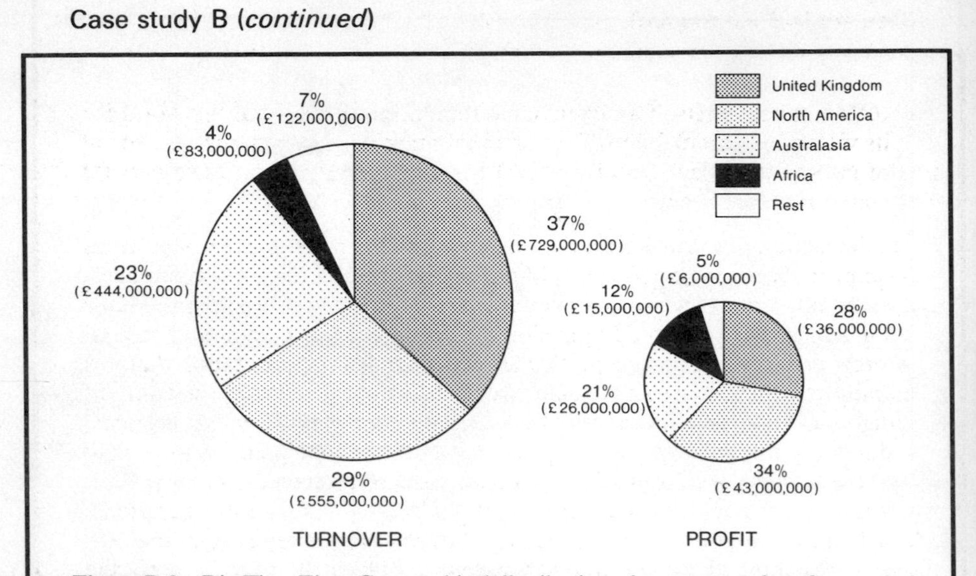

Figure B.2 Rio Tino-Zinc: Geographical distribution of turnover and profits, January–June 1985
Source: Rio Tinto-Zinc, *Annual Report 1985*

Table B.1 Rio Tinto-Zinc: corporate structure

Company name	*RTZ shareholding (% total)*	*Contribution to net attributable profits (1984) (£ million)*	*Main location of activities*
RTZ Borax	100.0	89.7	USA and Europe
RTZ Cement	100.0	15.73	UK
RTZ Metals	100.0	13.0	Europe and Middle East
RTZ Oil and Gas	100.0	13.2	North Sea
RTZ Pillar	100.0	54.3	UK,. North America, Australia
CRA	52.9	29.9	Australia, Papua New Guinea
Rio Algom	52.8	22.2	Canada
Rio Tinto South Africa	100.0	9.8	South Africa
Rio Tinto Zimbabwe	58.4	1.7	Zimbabwe
Rossing Uranium	46.5	7.3	Namibia
Enterprise Oil	29.8	6.9	North Sea
RTZ do Brazil	100.0	NA	Brazil

Source: Rio Tinto-Zinc, *Annual Report and Accounts 1984*

Case study B (*continued*)

> RTZ has earned itself an unenviable reputation throughout the world for its violations of indigenous land, manipulation of governments, disavowal of international law, implicit support for repugnant regimes and general contempt for its critics.

The formation of Partizans reflects the widespread criticism RTZ's operations have produced over the years. Without doubt some of this criticism reflects a wider anti-MNC sentiment, but certain aspects of the way RTZ operates have been subject to widespread opposition which is not just confined to vested-interest pressure groups. In particular, this includes the accusation that RTZ manipulates governments to gain access to mineral deposits in lands of indigenous but powerless peoples such as the Australian Aborigines, Canadian Indians and the black population of Namibia. The royalties paid to these peoples are claimed to be derisory and other benefits non-existent. RTZ is also attacked for the environmental impact of many of its operations, which are claimed to be widespread and adverse. Opponents claim that RTZ only cares about the environment where it is obliged to do so – rarely the case in the Third World. Finally, RTZ's involvement in the Rossing Uranium complex in Namibia is criticized for two main reasons. First, wage rates and work conditions are claimed to be deplorable, even by South African standards; second, RTZ's active support for the illegal and repressive South African occupation of Namibia inevitably draws the company into the wider anti-apartheid struggle.

Which of these pictures is true, that of the benign, socially responsible and rational business or that of the rapacious exploiter of peoples and resources, is hard to tell. It must be remembered that MNCS such as RTZ are not charitable organizations. They exist to make profits, not to provide a social service. If this is wrong, then the fault lies not with individual companies, but with the economic system in which they operate. Despite this, the evidence against RTZ suggests that it is far from blameless; that at times it has at best applied the letter, rather than the spirit, of the law and used its enormous power to guarantee operating conditions highly favourable to it. In the Third World the scope for it to do this is greater, and its impact more severe.

The impact of minerals mining on the Third World

The experience of many Third World minerals producers has not been a happy one. For most, being a mineral exporter has not led to a wider and sustainable process of economic development, and there is a strong line of argument which claims that the overall impact has been negative, rather than positive. It is argued that, in effect, mining operations in Third World countries are enclaves which have few links to the rest of the economy. There are, of course, exceptions to this, the capital-surplus oil exporters (in particular the Middle Eastern states) being the most important. There is also a point of view which holds that, whilst mining has not generated a wider development process, it does have substantial economic benefits for the countries involved. In particular it provides employment, infrastructure and vital foreign exchange for economies chronically short of all three.

It is clear from the above paragraph that there is some debate concerning the impact of mining on the Third World, one which is paralleled in many other spheres of development. Broadly speaking, the debate is between, first, those who argue that the structure of the global economy is such that the Third World is inevitably in a subordinate position, and that whatever form of economic integration into the global system occurs is inevitably unequal and exploitative, and, second, those who see economic interactions as essential to both the First and the Third World, because trade and the operation of markets are either mutually beneficial or can be made so by the actions of governments and international institutions. In other words, the debate over the impact of mineral production on Third World countries cannot be separated from a wider context of the general development debate. There is not the space to go into this debate in any detail here, but the student of First World–Third World links must bear it in mind when considering any particular aspect of those links (for further discussion of these points see the volumes by John Cole and Stephen Williams in this series).

Many Third World mineral producers are in effect caught in a trap. They are locked into an economic structure which depends upon exporting minerals, but this economic structure does not generate a wider and sustained pattern of economic development which permits them to break out of the cycle of poverty which is their lot. There are a number of reasons for this.

First, as we have seen, mining activities are in many cases owned and controlled by foreign-owned companies. This means that profits are taken out of the country and many economic benefits go to the MNC, not the host country. The control which MNCs exert over exploration, production and marketing makes it very difficult for the host countries to change this relationship.

Second, most stages in the full production process are located outside the country in which the minerals are mined, so employment, profits and other benefits from refining, producing finished goods and so on do not go to the host country.

Third, the long-term deterioration of the price of most minerals means that the Third World producers have in effect to run faster to stand still – they must expand their production and exports if they are to maintain the real level of their main source of foreign-exchange earnings. Most mineral exporters have not been able to do this, with the result that they get tied into an economic structure which requires foreign exchange and are then unable to sustain the pattern of investment and expenditure to which they become committed. Violent short-term price fluctuations make these impacts worse.

Fourth, minerals production is usually highly capital-intensive (generating few local jobs), is located in remote regions and has very few links with other sectors of the economy. In particular, few ancillary activities are generated, as most equipment is imported and no local inputs are required. As a result, minerals production forms a foreign-oriented 'enclave' which sits in, but is not really part of, the national economy and which generates few 'multiplier effects' – economic growth in other sectors of the economy.

Fifth, mining activities are highly concentrated, and whatever income and economic benefits they do generate in the host countries typically go to only a small section of the population. In particular, the urban élite frequently receive massive incomes, both legitimately and illicitly. What is left also goes to supporting state employees, developing urban infrastructure or paying for the armed forces. As such, income from mining frequently widens inequalities between rich and poor, urban and rural, and funds the system which creates and maintains these inequalities. Because they derive income and power from the activities of MNCs, all too often the ruling élites in mineral-producing Third World countries do little or nothing to protect the environment or the interests of local workers in the mining industries. A result of this is that MNCs are able to cut production costs by paying low wages, forcing workers to live and work in appalling conditions and paying no attention to pollution and environmental destruction (see the volume by Avijit Gupta in this series).

MNCs are also able to override the land and resource rights of local people in areas containing mineral deposits; stories of communities in remote areas being evicted (or worse) are common. Amazonian Indians, Australian Aborigines, the people of Namibia and many others have had their lands, their livelihoods and even their lives destroyed to enable mining companies to exploit mineral deposits.

Finally, because mining is by definition exploiting non-renewable resources, the exploitation of mineral deposits by export-orientated enclave production now precludes their future use to contribute to a fuller development process which retains the benefits of mineral production within the local economy and uses mining to generate growth and employment in other sectors of the economy. As such, mining without development at present means the loss of development potential in the future.

The factors listed above are all negative, but mining can, and does, have positive implications for the development prospects of Third World producers. It does provide scarce foreign exchange, generate government revenue through taxes and royalties, develop vital infrastructure (in particular transport facilities), generate some employment and result in the development of technical and managerial skills in the labour force. All of these benefits can be significant, but similarly all help to lock the country into a cycle of dependence on the mining industry. Once such reliance is acquired, the welfare of the state is locked into the health of the mining sector.

The response of Third World governments

The problems of dependence upon an externally controlled mining sector have long been recognized. In other words, it is not mineral mining itself which is problematic, but rather the form of organization and control which has characterized mineral mining in the Third World. This idea is the key to understanding the issues set out here. From the early 1960s onwards, changing the pattern of organization and control became a priority in many Third World countries. This was for a combination of economic reasons (attempting to keep more of the benefits of mineral exploitation within the national economy) and political imperative (everywhere MNCs had become symbols of neo-colonialism and seizing control of national resources from them a battle associated with national identity and self-determination). This self-assertion by Third World governments has led to some important changes in the way the mining industry operates in the past 25 years or so and has produced responses by MNCs and First World governments which have tried (with some success) to negate their effectiveness. We shall finish this section by briefly looking at these sets of responses.

Four types of policy have been introduced by various Third World governments to attempt to change the conditions under which the mining industry operates. The first is the renegotiation of the terms under which MNCs operate. The objective is to increase government revenue by introducing higher royalty payments, taxes and exchange control regulations. These policies have had some success for certain minerals, but the general effect has been to slow down the decrease in revenue which falling prices bring, rather than a substantial increase in revenue which provides the capital for a broader pattern of investment and development (the exception to this is oil).

Second, measures have been taken to try to dilute the enclave nature of MNC-operated mining activities. In particular, the aim has been to encourage links to other sectors of the economy, both by attempts to increase the proportion of locally produced inputs and by expanding local processing, refining and fabrication. These efforts have brought, at best, marginal success, with significant increases in such linkages largely confined to countries such as India and Brazil which have a broad industrial base and significant levels of internal demand for mineral products.

Third, since the 1960s state intervention has taken a more direct form, with the establishment of state mining corporations. These corporations have frequently been associated with the nationalization of foreign-owned mining activities; a policy which has been associated with nationalist politics and anti-MNC sentiment, and which has often brought an equally hostile reaction from the MNCs and First World governments. We shall return to this issue below.

The fourth policy Third World producers have pursued has been attempts to establish producer 'cartels' – associations of producing nations which try to agree policies that fix production quotas, raise the international price of the mineral concerned and so on. The best known, and by far the most successful, example of such a cartel is OPEC, but there have been similar attempts to establish such cartels for most major minerals. Most such attempts have failed in their efforts.

The effectiveness of these policies has varied from mineral to mineral and from country to country. Overall, the past 25 years or so have seen some erosion of the dominance of MNCs and a greater allocation of the benefits from minerals production to the Third World, but this has benefited some countries more than others and the fundamental structure of the world's minerals industry has not changed. Indeed, in the 1980s many of the gains made during the previous decade have been eroded, in particular by the decline of OPEC and a drop in oil prices in early 1986 which was almost as dramatic as the price increases of the 1970s. This section will continue with a brief look at why these attempts to change the conditions under which Third World countries participate in the international minerals industry have largely proved ineffective.

Minerals mining in the 1980s

Nationalization and the establishment of state production corporations have taken place in many countries. Where the nationalization took place without compensation, as in Chile in 1971, MNCs and First World governments have combined to respond. In the case of Chile, where the Allende regime nationalized the copper holdings of American MNCs, such as Anaconda and the Kennecott Copper Corporation, the response was in the form of a CIA-inspired coup and the installation of a military junta which promptly paid compensation to the MNCs.

In most cases the nationalization has been negotiated, and the compensation paid has proved expensive for the governments concerned. By the mid

1980s a significant proportion of minerals mining (but not processing, transportation and so on) was in the hands of state corporations. Superficially this seems a success story for the Third World, but in practice many MNCs have welcomed this development. The new state corporations are still dependent upon MNCs for key technology and expertise, and local control of the actual extraction process does not mean that they can opt out of the international economy. They are still totally dependent upon existing institutions to transport and market their output, and still subject to the vagaries of international mineral prices.

In addition, the new state corporations now have to bear the risks involved in mineral exploitation and production, and their ability to do so is much more limited than that of MNCs who have access to a wider range of capital resources and can spread their risks over different regions and activities. As such, most state mining corporations have only limited room to manoeuvre, and MNCs are able to negotiate joint agreements which are very favourable to them. This removes many of the uncertainties associated with large investments in the Third World. In consequence, the emergence of state corporations does not represent a wresting of control from the hands of MNCs. It rather reflects an urgent desire by Third World governments to maintain production (and foreign-exchange earnings), and a willingness to bear risks to achieve this. The MNCs may lose some profit margin from this, but this loss is outweighed by the benefits of reduced risks and greater flexibility in investment commitments. State corporations have consequently not redressed the imbalance, they have only done something to prevent this imbalance becoming greater.

Producer cartels have an extremely mixed track record. The outstanding success is OPEC, which succeeded in breaking the artifically low price of oil maintained by the First World (and especially the USA) and the oil MNCs, and in introducing two massive price rises over the winters of 1973/4 and 1979/80. There is not the room to tell the full story here (see Odell 1986 for the best account of events), but OPEC's success reflects the unique position of oil in the world's economy and owes more to political than economic forces. The success has also been short-lived. The immediate effect of the first oil crisis in 1973 was that the MNCs and the First World did everything they could to develop non-OPEC sources of oil, such as the North Sea and Alaska. In this they have been singularly successful. By 1985 OPEC was exporting less than half the amount of oil it had exported in 1973, and its market share had fallen to about a third of the world's total market. Falling demand for oil in the 1980s (caused by conservation measures and economic recession) and expanded non-OPEC supplies led to the collapse of oil prices in 1986, and with it the effective end of OPEC's power.

Even the brief success of OPEC has been unique. Far more typical of the fate of producer cartels has been that of CIPEC (the Intergovernmental

Council of Copper Exporting Countries). Formed in 1967 by the four major copper exporters (Chile, Peru, Zaire and Zambia), which in 1975 accounted for 65 per cent of world primary copper exports, CIPEC's aim was to increase the real income of its members. This was in response to the USA's policy of intervention to hold copper prices down. These increases were to be achieved by price and production agreements, both between CIPEC members and, where possible, with individual major copper consumers. Where OPEC succeeded, CIPEC failed. The organization tried to halt the fall in copper prices in 1974 and 1975 by reducing ouput, but with no success. Mikesell (1980) explains why. Because of the existence of alternative copper supplies and substitutes for copper, CIPEC members would have to reduce their output by over 2 per cent to increase the world price by 1 per cent – clearly a recipe for disaster. In addition, the copper-producing MNCs redirected their investments to non-CIPEC countries, leading to the further erosion of the market dominance of the CIPEC members.

The story of CIPEC, the copper cartel, has many parallels, for example in bauxite and iron ore, all of which have been as unsuccessful. One alternative to producer cartels is exemplified by the International Tin Council which regulated a commodity agreement between exporting and importers, but, as case study A has shown, the early successes of this approach have also come to nought.

Attempts by Third World mineral producers to improve their position in the world's minerals markets, so that their vital resources can form a broader base for development, have consequently largely proved unsuccessful. This reflects the structure of mineral industries (which involve much more than just mining) and, crucially, the ways in which this structure is controlled. It is these control mechanisms, in mining as in other spheres of economic activity, which create many of the development problems facing the Third World today. While they exist, these problems will remain; their form may alter, but their fundamental nature is the same.

Key ideas

1. Multinational corporations dominate the world's mineral industry. This reflects a process of capital concentration.
2. Mining is a risky business. Minimizing risk to massive investments is the main concern of mining multinational corporations.
3. Mining in the Third World is usually an enclave industry with few links to the rest of the economy.
4. Third World producers have tried to seize greater control of their mining industries, but these efforts have mostly failed.

Part II
Energy, survival and development

4
Energy crises in the Third World

Of all issues associated with resource development and exploitation, energy has attracted the most attention in recent years. It is also the one about which there are the most preconceptions and misconceptions. Ever since the dramatic increases in the price of oil over the winter of 1973/4 there has been widespread discussion of various forms of energy crises and gaps, a debate which reflects the traumatic impact of the oil price rises of that winter (and the similarly steep increases of 1979/80) on the world economy. It has also influenced the many projections of future price rises and oil-reserve depletion rates which have been made since 1974.

Since the early years of the 1980s such concerns have diminished to some extent with the emergence of an oil glut and in particular with the fall in oil prices over the winter of 1985/6, a fall which was almost as dramatic as earlier price increases. Despite this the events of the decade following the first 'oil crisis' in 1973/4 have undoubtedly focused attention on energy as an issue, attention compounded by the recent disaster at the Chernobyl nuclear power plant in the Soviet Union, by previously undisclosed reports of near disasters in the British and French nuclear industry, and by the subsequent debate on the merits of nuclear power.

The problems which confront less developed countries (LDCs) concerning energy supplies are diverse and complex. There is inevitably a degree of variation in the form they take in different places, but there are common difficulties which confront most, if not all, LDCs which are not major oil exporters. There are a number of parallel dimensions to these problems, and we can separate them to help clarify the discussion which follows.

First, energy must be considered as a vital input into the process of production. Whatever is made, energy is necessary to make it. More specifically, fairly rapid increases in available energy supplies are an essential input into a process of economic development, whatever form such development takes. If such a development process includes any significant industrial and infrastructure development, then the energy requirements are even greater. Problems associated with fuel costs and continuity of energy supplies are having a major impact upon the pace and form of economic development throughout the Third World. This impact has been particularly acute since the second oil crisis of 1979/80. This dimension of Third World energy problems has been fairly widely recognized and can be considered as a *crisis of energy for development*.

The second dimension of energy issues is energy as an aspect of patterns of consumption. We all need and use energy on a day-to-day basis. Energy is a basic need; it is vital for survival, and there is increasing evidence that the ability of many communities to provide the energy for their basic survival needs is seriously under threat. This crisis is closely associated with poverty; it particularly threatens poor people and poor countries. This issue has only been recently recognized and still receives very little attention. It can be considered as a *crisis of energy for survival*.

Energy sources and balances

At this point some clarification of definitions is necessary. When we use a fuel (coal, wood, oil, etc.) we do not directly use the physical substance itself, but rather the *power* (i.e. the potential to do work) which is derived from the *energy* released during the consumption of the *fuel*. Energy can thus be derived from many sources, including some, such as solar, wind or hydropower, which are not physical fuels.

The implication of this is that we can satisfy any one particular energy need from a number of different energy sources, or in other words that there is a degree of *substitutability* between different fuels. In practice, however, such substitution possibilities are limited. For any particular application of energy (e.g. cooking), different sources of energy (e.g. wood, gas) will be harnessed through different forms of technology (e.g. open fire, gas stove), all of which have a different range of economic costs and resource constraints. The result is the world as we know it: particular sets of fuel – technology combinations associated with specific end-uses and limited substitution possibilities in any one case.

Within the context of Third World energy, we can identify three broad categories of energy sources:

1 *Commercial* or modern fuels: oil (in its various forms), coal, gas and

electricity (including that generated in large-scale hydroelectric power schemes). These fuels have a commercial value, are frequently imported (this is especially true for oil), have only been used widely in comparatively recent times and are the fuels most closely associated with the crisis of energy for development.

2 *Non-commercial*, traditional or biomass fuels: wood, charcoal and crop, and animal residues. These fuels are usually non-commodified (i.e. are not bought and sold), may be gathered freely from the local environment, have been used extensively for many thousands of years and are the fuels most closely associated with the crisis of energy for survival.

3 *New and renewable* (alternative) sources of energy: wind, wave, solar, mini-hydro, ethanol, biogas, tidal and so on. These sources of energy have been widely advocated as 'appropriate' alternatives to both fossil and biomass fuels but have made very little impact upon Third World energy consumption. Their potential appears to be limited and, where it does exist, is associated with particular forms of development.

Conservation, in one sense, could also be considered an energy 'resource', because using energy more efficiently produces the same work with less fuel. Conservation is certainly a policy option, and research and development into energy-efficient technologies for the Third World is one of the issues examined below.

The role and relative importance of these various energy sources are reflected in national *energy balances*. An energy balance is a table which shows, in a standard energy unit, the quantities of different fuels produced and used in different sectors of the economy. Some examples of summary energy demand balances are shown in table 4.1.

Case study C

Energy in Indonesia

A brief look at the case of Indonesia will help illustrate the point that energy problems are not created by physical-resource scarcity at a national level; they rather reflect economic constraints which retard resource development and the limits of space where energy supplies are separated from energy demand. Indonesia contains most energy resources in abundance but also faces most of the energy problems which characterize the Third World. Despite her size and revenues from oil exports, Indonesia lacks the financial capital, human resources and physical infrastructure to harness fully her energy-resource potential. The great distances between potential resources and centres of demand compound these constraints, as does the low income

Case study C (*continued*)

Table C.1 Indonesia: energy demand by sector (1978,'000 TOE)

	Petroleum	*Biomass*	*Others*	*Total*	*(%)*
Household	4,380	35,471	83	39,934	(76)
Industry	2,930	259	1,496	4,685	(9)
Transport	4,535	—	21	4,556	(9)
Power generation	1,150	—	532	1,682	(3)
Others	1,683	98	—	1,781	(3)
Total	14,678 (28%)	35,828 (68%)	2,132 (4%)	52,638	(100)

Source: derived from Government of Indonesia Directorate General for Power (1981) *Energy Planning for Development in Indonesia*, 59

of many of the people who face the severest energy problems.

Table C.1 shows the pattern of energy demand in Indonesia for 1978 (the pattern has changed little since then). Although Indonesia contains massive reserves of oil and other fossil fuels (as well as great potential for many forms of new and renewable sources of energy), biomass energy (mostly wood) is still by far the most widely used fuel. Similarly, despite economic growth, the household sector is the main user of fuels. In urban areas, a wide variety of fuels is used, and kerosene (which was heavily subsidized until recently) is particularly important both for lighting and cooking. Wood is still important in the cities, however, and the impact of Jakarta's demand is felt far and wide.

Rural households use some kerosene and electricity (where it is available) for lighting and, occasionally, cooking, but mostly rely heavily on wood for their energy needs. Almost 95 per cent of biomass fuel demand in Indonesia is found in rural areas. This constitutes about 130 million cubic metres of wood and agricultural wastes a year, an enormous quantity of material even for a biomass-resource-rich country such as Indonesia. A sharp distinction can be drawn between densely populated islands (principally Java), where the rural population is faced with serious problems of access to energy resources and few forests remain, and more sparsely populated regions such as most of Kalimantan and Irian Jaya, where there is no significant energy problem and much of the land is covered by lush rain forests.

In Indonesia non-household energy demand is concentrated in transportation and industry. The important transport sector relies almost entirely on oil. Demand has grown rapidly. The number of motorized vehicles in Indonesia grew from less than 600,000 in 1967 to 3,300,000 in 1979. Industrial use of energy in Indonesia is more varied, with oil, coal, natural gas, electricity and fuelwood all contributing. Growth of demand for energy in the

Case study C (*continued*)

industrial sector has been rapid, averaging 12.9 per cent in the 1970s. Around 85 per cent of manufacturing enterprises in Indonesia are located in Java, principally in and around Jakarta.

Figures C.1 and C.2 show the distribution of Indonesia's energy resources. As figure C.1 reveals, large areas of Indonesia have abundant biomass resources and few fuelwood problems. Indonesia's total forest area in 1980 covered 113,895,000 hectres, or 59 per cent of the total land area.

In the less densely inhabited areas, wood is used almost exclusively and is freely gathered from predominantly forest areas. There only exist difficulties of access in very localized areas such as, for example, around some agricultural resettlement schemes. In Indonesia as a whole the total forest area could easily sustain the fuelwood requirements of the total population for the foreseeable future, but the different locations of demand and supply and the prohibitive cost of transporting the fuel to the people mean that this theoretical potential has little practical relevance.

For the population of densely settled areas of Indonesia, the main source of biomass fuels is not forest areas, but is material extracted from agricultural land. A number of detailed case studies in different parts of Java have shown the importance of trees outside the forest as a source of fuel, the problems which people in densely populated areas face in meeting their energy needs and the impact of the urban fuelwood market on rural fuel supplies. In the worst-hit areas crop residues are increasingly important as a fuel source.

Indonesia has considerable deposits of oil, coal and natural gas. The resources are located in many parts of Indonesia (figure C.2), but the eastern part of Kalimantan and Sumatra are paticularly important areas of current production. Proven oil reserves in the early 1980s were about the equivalent of 20 years' output. Estimates of undiscovered or unproven resources vary widely. Low estimates are in the order of 1.4 billion metric tons (or about the equivalent of proven reserves), with some estimates as much as four times this total. The higher figures suggest that current production levels could be maintained until well into the second half of the next century.

Indonesia has vast reserves of natural gas. The proven reserves of 772 million TCE (metric tons of coal equivalent) have expanded considerably with the discovery and exploration of the giant Natuna field in the South China Sea (figure C.2), which has estimated reserves of 35 trillion cubic feet of gas, or half of current reserves. As with oil, it is estimated that there are

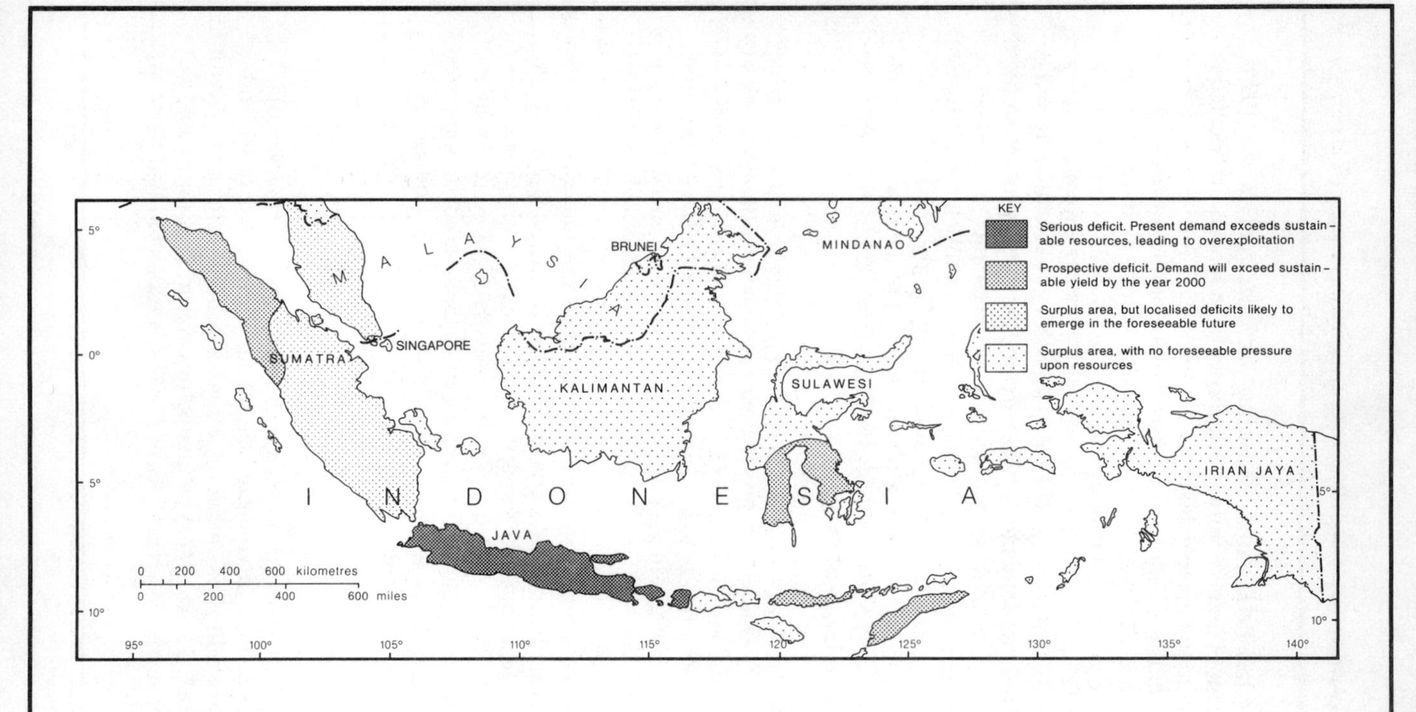

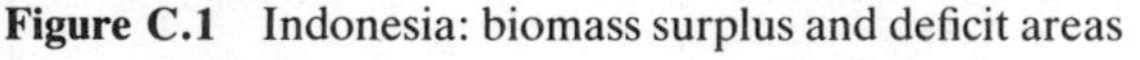
Figure C.1 Indonesia: biomass surplus and deficit areas

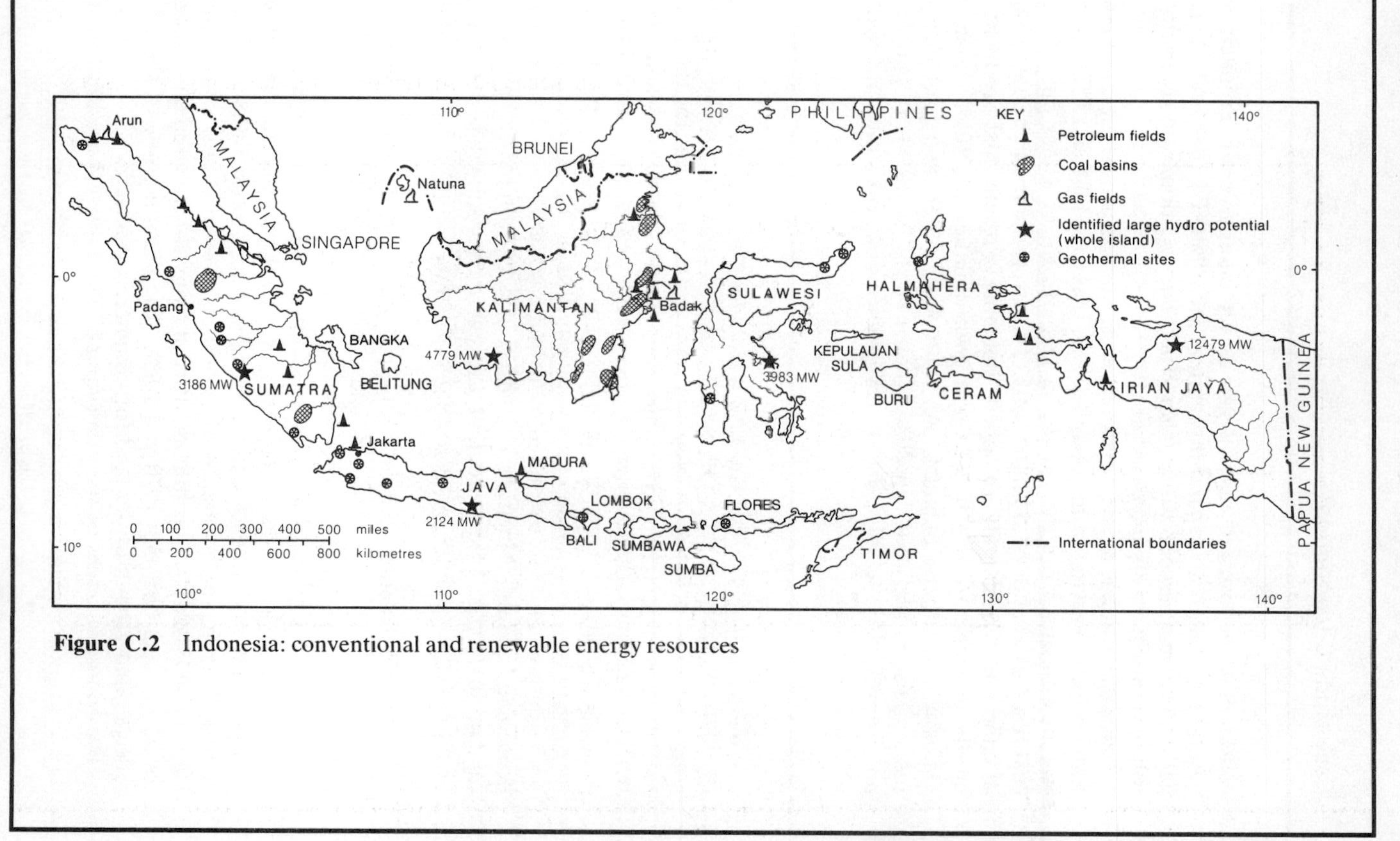

Figure C.2 Indonesia: conventional and renewable energy resources

Case study C (*continued*)

considerable deposits of natural gas as yet undiscovered. Production of natural gas is likely to expand throughout the 1980s. By 1990 revenue from gas exports may exceed those from crude oil, and domestic consumption will have undoubtedly increased greatly.

Indonesia also has massive coal reserves. Their exact extent is unknown, with estimates varying from about 3 billion tons to as high as 15 billion tons. Recent estimates from the Indonesia Department of Mines put mineable coal reserves at about 2.6 billion tons. These represent an energy resource of enormous, and largely untapped, potential. Present coal production is mostly confined to two small, state-owned mines at Bukit Asam and Ombilin in southern Sumatra. As with wood, the main problem with Indonesia's gas and coal deposits is that they are found where people are not. They do little to solve the energy problems of Java's millions, and are unlikely to do more in the future.

The figures in these balances give a good indication of the pattern of use of different fuels in different sectors of the economy of the countries concerned. The balance for Nepal is representative of that for the world's very poorest countries. The table reveals the absolute dominance of biomass fuels amongst the energy sources used, and of the household sector amongst the areas of energy demand. The consumption of commercial fuels, at 10.9 kg oil equivalent per capita in 1980, is largely confined to what road transport there is, to limited industrial production and to kerosene for household lighting. Much of it is used in and around Kathmandu, the capital. In Nepal, 94 per cent of the energy consumed is provided by biomass fuels (mainly wood) used for cooking the food of Nepal's population.

The energy balance of Brazil is representative of the fuel consumption patterns of the better-off LDCs. Biomass fuels are far less important (but are still of significance) and the consumption of oil and other commercial fuels is far greater. This reflects the development process Brazil has experienced in recent decades, and in particular the rapid growth of urbanization, industry, transport and other intensive users of commercial energy. In recent decades biomass fuels have declined as a proportion of total energy consumption, but even in Brazil the absolute quantity of biomass fuels used has increased (by 25 per cent between 1970 and 1980). The declining role of these fuels reflects rapid growth of demand for commercial fuels.

The energy balances of the other countries in table 4.1 are illustrative of

Table 4.1 Energy demand balances for selected countries

	1,000 metric tons oil equivalent			
	Biomass	*Petroleum*	*Others*	*Total (%)*
Nepal (1980/1)				
Household	3,089.3	30.4	6.6	3,126.3 (95.1)
Industrial	2.1	4.8	39.4	65.3 (2.0)
Transport	—	69.1	1.2	70.3 (2.1)
Others	9.0	14.3	2.8	26.1 (0.8)
Total	3,119.4 (94.9%)	118.6 (3.6%)	50.0 (1.5%)	3,288.0 (100)
Kenya (1980)				
Household	4,058.6	173.2	45.8	4,277.6 (60)
Industrial	1,127.7	419.8	68.2	1,615.7 (22)
Transport	—	997.3	1.1	998.4 (14)
Others	43.8	188.1	41.8	273.7 (4)
Total	5,230.1 (73%)	1,778.4 (25%)	156.9 (2%)	7,165.4 (100)
Grenada (1982)				
Household	13.6	4.6	1	19.2 (56)
Industrial	—	3.8	0.2	4.0 (12)
Transport	—	8.1		8.1 (24)
Others	0.9	1.3	0.8	3.0 (9)
Total	14.5 (42%)	17.8 (52%)	2.0 (6%)	34.3 (100)
Brazil (1980)				
Household	12,071	254	4,960	17,285 (19)
Industrial	15,432	15,766	11,068	42,266 (46)
Transport	1,501	24,341	94	25,936 (28)
Others	508	3,177	2,339	6,024 (7)
Total	29,511 (32%)	43,538 (48%)	18,461 (20%)	91,510 (100)
Sri Lanka (1982)				
Household	40	185	2,925	3,150 (66)
Industrial	70	276	635	981 (20)
Transport	—	592	—	592 (12)
Others	35	47	3	85 (2)
Total	145 (3%)	1,100 (23%)	3,563 (74%)	4,808 (100)

LDCS with levels of development somewhere between very poor countries, such as Nepal, and more prosperous LDCS such as Brazil. The balances show that biomass fuels are still dominant, but oil is of increasing importance. This reflects the growth of energy use outside the household sector, with transport being particularly important and, in some cases, with industry also emerging as a significant energy user.

The energy balances discussed here indicate the relative importance of different sources of energy, with particular fuels clearly being associated with different forms of energy use and, in consequence, energy problem. This underlies the energy 'crises' identified above. This chapter is concluded by discussing this dual crisis in greater detail.

Energy for development

In comparison to western nations, Third World countries are characterized by very low levels of energy consumption. There is a great deal of variation in this (see table 4.2), with both the types and quantities of fuels used showing some relationship to the level of economic development as measured by per capita GNP. This relationship is far from perfect, however, and it must not be taken as some form of 'law'. The level of economic development of a country is the most important factor influencing the pattern of energy consumption, but many other factors also affect this pattern and create the varied trends which table 4.2 shows.

Table 4.2 Energy and economic indicators for selected countries

Country	*Per capita* GNP *(US$ 1984)*	*Per capita commercial energy consumption (gigajoules 1983)*	*Fuelwood as % of total energy consumption*	*Energy imports as % merchandise exports (1984)*
Ethiopia	110	1	93	48
Bangladesh	130	0.5	80	20
Burkina Faso	160	1	94	86
Nepal	160	0.5	97	49
India	260	7	36	59
Madagascar	260	2	80	32
Benin	270	1	86	53
Kenya	310	3	70	51
Sierra Leone	310	2	76	63
Sri Lanka	360	4	74	33
Zambia	470	10	46	5
Honduras	700	7	45	28
Nigeria	730	6	82	3
Nicaragua	860	9	25	46
Thailand	860	10	63	33
Peru	1,000	18	34	3
Brazil	1,720	19	33	30
Malaysia	1,980	26	8	12
South Korea	2,110	43	5	25
United Kingdom	8,570	137	—	15
United States	15,390	273	—	29

The question of the availability of energy to fuel the process of development is principally associated with fossil fuels, and in particular imported oil. Put simply, the pattern of development found in most LDCS and the relative trends in prices of goods on the international markets (the terms of trade) mean that LDCS are forced to run increasingly quickly to stand still. To develop they must import fuel, but the very cost of that fuel retards their development prospects. Why is this the case?

This pattern characterizes LDCs because the process of economic development does not mean just general growth of the pre-existing economic structure. During economic development some sectors of the economy grow far more rapidly than others, and these leading sectors are generally the more intensive users of energy, such as industry and transport (plate 4.1). Such development is also generally associated with a transformation in the pattern of energy consumption, as the society becomes increasingly urbanized and the urban population consumes more energy (and in particular commercial energy) than their rural counterparts. As such, LDCs undergo what is called an *energy transition*.

Plate 4.1 Energy for Development: Trucks in Bangkok

This effect is particularly associated with the growth of large-scale industry and the development of the transport sector. In both cases, in order for growth to occur, greatly increased supplies of energy are necessary. For LDCs in the contemporary world there is little hope of any sustained economic transformation if these sectors do not grow. In consequence, in most LDCs the growth of demand for commercial fuels is more rapid than the growth of the economy as a whole. This is very important, as it means that any detrimental impact of increasing commercial energy consumption will be compounded over time and may itself become a retarding factor upon economic development.

The key to this is oil. Most LDCs are heavily oil-dependent for commercial fuel supplies, and this oil is generally imported. In such cases, the cost of oil imports is a major drain upon their foreign-exchange earnings. The impact can be expressed as a percentage of exports, as shown in table 4.2. These figures have increased over time and by the early 1980s were at a level where oil imports accounted for 50 per cent or more of the exports of many LDCs. This increase in the cost of oil imports over time has not occurred gradually. It reflects the two dramatic rises in the price of oil in the so-called oil crises of 1973/4 and 1979/80.

The impact of these rapid price increases upon the economies of LDCs was traumatic, and many have yet to fully recover from the economic dislocation they caused. The official price of oil on the international market rose from $3.01 per barrel in October 1973 to $11.65 per barrel in January 1974, and again from $13.34 in January 1979 to $18.00 in October 1979, $26.00 in January 1980 and $32.00 in January 1981. (A note of caution is needed here. These prices are for 'Saudi light'. In 1979/80 the Saudis tried to slow the price rise. In late 1980 the price of oil on the Rotterdam spot market rose to over $40, and many countries increased their prices more sharply in 1979 than the Saudis.) The effects of the second price rise, in 1979/80, on the economies of many LDCs was particularly severe, as it was followed by a period of general economic recession and falling commodity prices (see figure 4.1). This meant that export earnings declined at a time when the need for foreign exchange to pay for oil imports (as well as debt servicing and other vital imports) was increasing rapidly. The effects of this were that many LDCs (and in particular the very poor countries of Africa, Asia and elsewhere) had problems maintaining existing levels of oil imports and found it impossible to consider the increased oil imports necessary for any sustained economic development.

The crisis of energy for development is, consequently, centred on the cost of the energy which literally powers any process of development. This problem relates closely to the issues raised in section 1 of this book, in which international markets for and control of mineral resources were discussed. Most Third World countries have little, if any, scope for influencing these markets and are subject to the fluctuations which characterize them. This lies at the heart of the crisis of energy for development.

Energy for survival

The second or hidden energy crisis in the Third World is that which faces significant sections of its population. The welfare of these people is increasingly jeopardized by the difficulty they encounter in catering for their

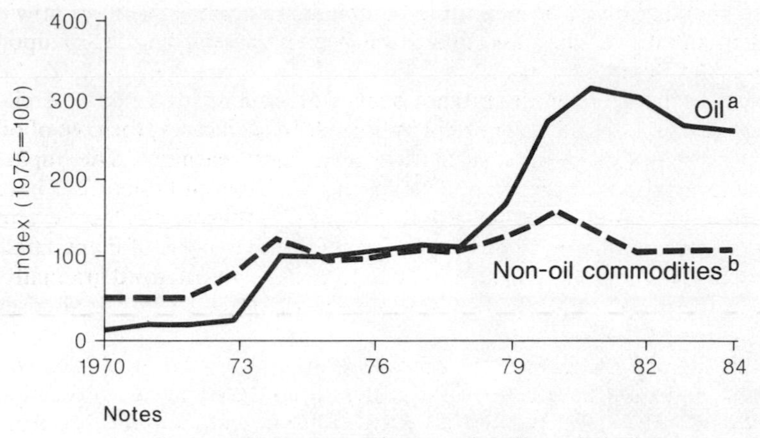

Notes

a. Average dollar price of internationally traded oil

b. Average dollar price of 33 primary commodities, weighted by each commodity's share in developing countries' exports

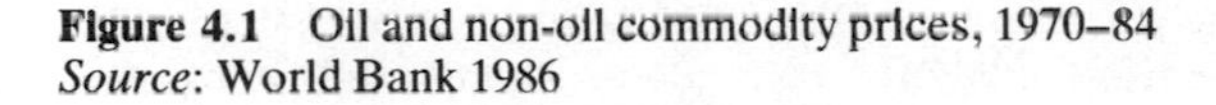

Figure 4.1 Oil and non-oil commodity prices, 1970–84
Source: World Bank 1986

basic energy needs. We all need energy of one form or another for day-to-day life, for cooking, lighting, heating and so on. Energy can consequently be considered to be a basic need, along with food, water, shelter and others.

This aspect of Third World energy use has received far less attention than the issues associated with commercial energy supplies but is just as critical for the future prospects of these countries. The fuels used to satisfy these basic needs vary from place to place, but, as the energy balances in table 4.1 show, the main energy source for Third World households is what is known as biomass, or traditional, fuels. This is particularly true for cooking (plate 4.2) which is the largest user of energy for most Third World households.

The scale of these hidden problems is difficult to evaluate, partly because they have been ignored for so long. But it also reflects the way in which the problems express themselves, for they rarely emerge in the form of an absolute lack of fuel. When wood becomes scarce, people are forced to spend a longer time and to go further to gather it, to switch to less favourable fuels such as agricultural residues, to start buying fuels, when previously they gathered them freely, or to adopt a range of other responses to the problem of scarcity. These difficulties rarely affect, and are often not noticed by, national governments or powerful commercial interests.

Few people paid any attention to this problem until the late 1970s. Since then, concern has grown, not least in major international agencies, such as

Plate 4.2 Fuelwood for cooking in Malawi

the World Bank and the UN Food and Agricultural Organization (FAO). There is now a tendency to over-state these problems in some cases. For example, a widely quoted study by the FAO in 1980 suggests that there are some 1.3 billion people (about 40 per cent of the Third World's population) living in fuelwood 'deficit' areas. This exaggerates the problem; things are bad, but not that bad. It also tends to miss the point. As with most other dimensions of extreme poverty, acute fuelwood problems are mostly highly localized, with the form they take varying from place to place. Whole regions and countries are only affected under the most extreme circumstances – such as in Lesotho, much of Nepal and across the Eastern Sahelian belt in Africa. Elsewhere in the Third World only some people are affected by *major* energy problems (though many more will experience minor difficulties). Unfortunately almost everywhere those worst affected are the poorest and least powerful sections of the community, with women in general and the land-poor and landless bearing the brunt of this problem, and with the more powerful local groups being the last to be affected.

Conclusions

The energy question in the Third World is a complex and important one. Energy problems are undermining development strategies and jeopardizing the survival prospects of the poorest and most vulnerable sections of the community in many parts of the Third World. This reflects competition for and conflicts over scarce resources both locally (in particular for fuelwood) and internationally (for oil and other fossil fuels). This idea of resource scarcity lies at the heart of such problems. Most fundamentally, scarcity occurs when demand for a resource is greater than the physical quantities available. In a commercial system this is reflected in the price of the resource, which rises as pressures upon the supplies increase. However, this relationship assumes a 'perfect' market; something rarely found anywhere, and certainly not typical of Third World energy markets.

Much of the scarcity of both commercial and biomass fuels in the Third World does not reflect absolute physical shortages of these fuels. Rather it reflects limited access to these fuels. At the national and international level this limited access is a consequence of economic weakness (and in particular foreign-exchange shortages), shortages of the capital and expertise necessary for the development of energy resources and infrastructure, and the ways in which international energy markets work (which are loaded against the interests of the Third World). At the local level access to biomass resources is limited by many factors (discussed in the following chapter). The basic point is that resource scarcity in the real world is a consequence not of the physical unavailability of resources, but of the way in which these resources are controlled.

It is the management of energy resources which limits their availability to the poor. As such, energy problems in the Third World reflect poverty in the Third World. They are just one aspect of the development problems which poor people and poor countries face. The root causes of energy problems will not be removed while poverty exists, but providing adequate and secure supplies of energy at a cost people can afford will itself be a major contribution to removing poverty. We will look at these issues in depth in the following two chapters.

Key ideas

1 The Third World faces a dual energy crisis: the crisis of energy for development and the crisis of energy for survival.
2 The cost of commercial fuels, which are essential for economic development, retards the development process.
3 The world's poor face a hidden energy crisis of fuelwood availability which threatens their survival.

5
Energy and the environment: how poor people cope

So far we have looked at minerals and at energy in terms of international economic trends and, for Third World countries, national development prospects. In other words, we have looked at the macro level, with the problems expressed in terms of control of international markets, commodity price movements, investment requirements, foreign-exchange availability and so on. These are very real problems; indeed they are the problems with which governments (as opposed to the people) are most concerned, as they directly affect the operation, policies and prospects of governments. These forces also affect many people in real ways, as they influence the income they receive and the price they pay for many vital goods and services. For the very poor, however, these problems are remote and secondary to the more immediate problems they face – what has been called the *survival crisis*. These problems are much more real, more immediate, more direct for the very poor in particular. This chapter examines the ways in which people obtain and use the fuels essential to their basic needs in rural areas and, more briefly, in towns and cities.

In chapter 4, it was noted that biomass fuels (wood, charcoal and agricultural residues) are by far the most important source of energy in much of the Third World, and that these fuels are predominantly used in the household sector, to meet people's needs for cooking, heating and so on. However, Third World populations do not use only these fuels. Many use the commercial fuels which are familiar to us: electricity, oil products such as kerosene, gas and so on. Their use is restricted, however, both by their cost (for most they are more expensive) and by their availability (in many places commercial fuels are difficult to obtain). Use is also restricted by the cost and availability of the appliances (especially cooking stoves) which are needed to make them useful – an open, three-stone fire costs nothing; an electric or gas stove needs a large lump-sum payment which poor people are unlikely to be able to save. In other words, patterns of household energy use vary according to both *location* (with in particular, differences between rural and urban areas) and *social class* (depending on the level and form of income, access to credit, control over resources and so on).

Energy in rural areas

Most of the Third World's poorest people live in rural areas, where biomass fuels are the main source of energy. They are used by all, poor and rich alike, but it is the poor who are hit first and hardest by shortages of such fuels. Shortages are not universal; in many regions even the poorest, most disadvantaged sections of the community have little or no problem obtaining adequate fuel supplies. In other words, biomass fuel problems are *specific to people in places*. They reflect the resource and socio-economic characteristics of specific localities and cannot be separated from other aspects of resource management and people's lives. What this means is that rural energy (and in particular biomass fuels) can be fully understood only as a component of an integrated rural production system. This concept is relatively new but now forms a framework within which both academics and development agencies, such as FAO, are approaching rural development issues in the Third World. An example from Kenya of how this approach can be used to understand rural fuelwood use is given in case study D.

Case study D

Understanding rural energy: woodfuel and trees in Kenya

Viewing fuelwood production and use in rural areas as a component of an integrated production system necessitates a detailed knowledge of specific localities if a real understanding of rural energy problems is to be achieved and effective planning interventions identified. Few rural energy studies provide this detailed picture. The Beijer Institute Kenya Fuelwood Cycle Project (KFC), and the Kenya Woodfuel Development Project (KWDP) which followed on from it, is one of the most complete and effective studies to date. The main conclusions of these programmes are summarized here, with the material presented drawn directly from project publications. More complete summaries are available in the edition of the journal *Ambio* listed at the end of this volume.

Case study D (*continued*)

The KFC began with a national survey of rural energy demand. Biomass fuels (mainly wood and charcoal) were found to provide the vast majority of rural households' energy use. National average household consumption of different fuels in 1981 was 4.75 metric tons of wood, 664 kg of charcoal and 52 litres of kerosene (which is used largely for lighting). Each person in Kenya consumes, on average, over a ton of wood (including wood used to make charcoal) per year for fuel alone; a figure not untypical of many parts of the Third World. Fuel consumption varies from area to area. Perhaps surprisingly, consumption was highest in the semi-arid zone (averaging 8.4 tonnes of wood and 570 kg of charcoal per annum) where the resource base is lowest and most fragile, but where very low population densities and highly mobile nomadic communities do not place severe pressures on these resources as long as traditional patterns of resource use are not disrupted. Lloyd Timberlake (1987) describes what happens when such disruption occurs. In northern Kenya nomadic Rendille people have been attracted by permanent water, disaster relief supplies, schools and services to settle in the town of Korr. This has both undermined the viability of their pastoral economy and led to a circle of environmental degradation extending 20 miles around the town as trees were felled for food and pastures overgrazed by livestock, a clear illustration of the fragile nature of production systems in semi-arid areas.

Despite this, fuelwood consumption is lower in high-potential areas where the landscape is one of permanent and highly intense arable farming. In Kenya, 80 per cent of the population live in the 20 per cent of the country which forms the highlands – the only areas where arable farming is really viable. Population densities are extremely high and pressures upon fuelwood resources severe. In arable areas the average consumption is 4.3 metric tons of wood and 380 kg of charcoal per household in the medium potential zone and 4.6 metric tons of wood and 786 kg of charcoal in the high potential zone. In these areas little communal land is left. This is particularly true in the best zones, where the landscape is one of large commercial plantations and *shambas* – the small, densely packed farms of Kenya's peasants. In these areas, woodfuel supplies come almost entirely from non-forest areas, with private farms the major source.

Consumption also varies by socio-economic groups. Income, household size, the cost – in terms of time or cash – of wood and charcoal and dietary patterns all influence the type and quantities of fuels used. This can be related to the relationship of individual households to the transition from

Case study D (*continued*)

precapitalist to capitalist models of production. Not surprisingly, the greater a household's access to land and cash income, the greater its energy consumption and the more secure its energy supply. This demonstrates the central point that energy problems are just one aspect of wider problems of poverty.

The KWDP produced a far more detailed analysis of the way fuelwood fits into the production and use of woody biomass in general, which itself is one facet of local production systems. Kakemega District, in Western Province, is an area of high ecological potential which contains population densities of up to 1000 persons per square kilometre. The pressures on fuel supplies and other uses of wood are acute. The KWDP revealed that trees are an integral part of rural production, with farmers consciously cultivating trees on their own farms. Figure D.1 shows the varied niches trees occupy, covering one third of the total farm area. These trees perform many functions, including providing a cash income from the sale of poles from the eucalyptus woodlot for construction. These trees are the main source of energy for the farm households. Surprisingly, the KWDP found that the coverage of trees on farms was highest in the most densely populated areas of Kakemega District. In these areas, as elsewhere, the varied products trees provide are a vital component of the household production, yielding a range of goods needed for the family's survival and cash income from the sale of wood and other tree products. The production system in Kakemega is a classic example of indigenous agroforestry, in which an integration of trees, bushes and ground crops permits farmers to maximize the total yield from their small patches of land to meet their needs for cash income and subsistence goods.

Despite the abundance of trees, however, many women in Kakemega report problems in meeting their fuel needs. This paradox reflects the different roles of men and women, with the women often denied access to wood to meet their role as energy providers by the men, who control the land and trees and who use the wood for other purposes. As such, the pattern of control of resources and competing demands for woody biomass restrict access to fuelwood supplies and lead to energy problems.

The KWDP concluded that the solution to the energy problems of areas such as Kakemega must come from the woody biomass cultivated and managed by farmers themselves. In areas such as this, there is no alternative to fuelwood, and the only viable source of supply is from the farms themselves. As such, solutions to rural energy problems must be integrated

Case study D (*continued*)

into the local production system. How to do this is the challenge, but such a solution will attack the root cause of the problem poor people face in rural areas of the Third World.

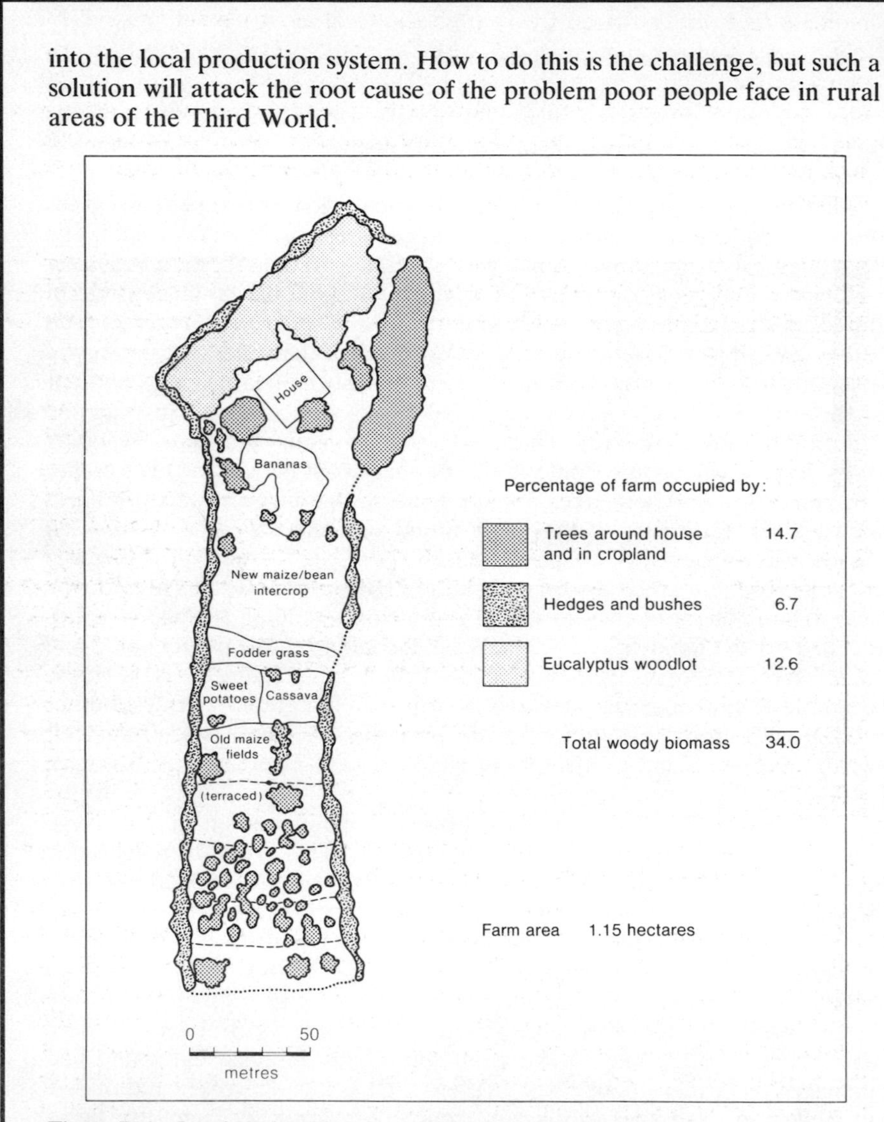

Figure D.1 Land use in a *shamba* in Kakemega District, Kenya
Source: Bradley *et al*. 1985

In rural areas fuelwood use can be understood only at a local level because it is produced and consumed at this level. In most rural areas, biomass fuels are gathered freely from the local environment (plate 5.1). They are a non-commercial good, with the costs associated with their use being derived from the time taken to collect them, from the impact on the local environment and from the 'opportunity cost' of their use (if biomass is used as a fuel it is not available for other vital uses). Non-monetary costs such as these are extremely difficult to measure and vary greatly from place to place.

Plate 5.1 Women gathering fuelwood in Zimbabwe

Collecting these fuels is women's work in most regions of the Third World (often with the help of their children). This gender division of labour is important, as it is women who experience fuelwood problems but men who control cash income and land resources and consequently control the solutions. The time it takes to collect fuels is time which is not available for the many other tasks women face in the Third World (see figure 5.1). Where fuelwood resources are abundant and freely available to the women in a community, these time costs are low. All too often this is not the case, however. One of the first signs of fuelwood problems is the gradual increase in the time it takes the women of a community to find the fuel essential to their households' survival. Eckholm *et al.* (1984) cite examples from a number of countries:

> In parts of rural Peru, women spend 10 per cent of their time gathering and cutting wood. In northern Ghana, one full day is required to gather three days' supply of wood. Women . . . may walk eight kilometres to their husbands' bush farms to collect headloads. A recent survey in rural Kenya found that some women spend up to 20–24 hours per week collecting fuelwood.

Productive work	**Household maintenance**
Preparing soil	Collecting water
Preparing seeds	Gathering firewood
Planting seeds	Collecting/preparing dung cakes
Transplanting rice/seedlings	Preparing vegetables/fish
Weeding/tending plants	Cooking meals
Irrigating	Preparing/cooking rice/bread
Applying fertilizers	Preparing milk products
Harvesting	Making/tending fire
Carrying crops	Washing pots and utensils
Threshing	Washing clothes
Winnowing	Cleaning stove
Storing	Cleaning house/compound
Husking	Sweeping/cleaning cowsheds
Milling	Sewing/mending clothes
Keeping vegetable garden	Feeding/washing/caring for children
Feeding livestock	Serving meals
Collecting eggs, milk, etc.	Carrying/nursing infants
	Caring for sick
	Supervising children/household workers

Figure 5.1 Women's work in a peasant household
(based on surveys in India and Bangladesh)

A major study of nine countries in Southern Africa by the ETC Foundation (1987) found increasing fuel-collection times in many areas across the region. The same is true for many regions in Asian countries such as Nepal (where deforestation makes journeys of several hours up mountain slopes increasingly common), Sri Lanka, Indonesia and Thailand. Agarwal (1986) has summarized data on the time taken to collect fuel from a number of studies (table 5.1). Increasing collection times place a great burden on the labour budget of the women involved, as they walk further and spend longer to gather their fuel. This reveals a basic feature of rural energy problems in the Third World. They rarely express themselves directly as an absolute lack of fuel. Only in the most extreme circumstances of famine and environmental disaster are people unable to cook their food. Rural energy problems express themselves indirectly, with increased collection times one of the key indicators of deteriorating fuel availability.

Table 5.1 Time taken and distance travelled for firewood collection

Place/date	*Firewood collection*	
	Time taken	*Distance travelled*
Nepal		
Tinan (hills) 1978	3 hours/day	—
Pangua (hills) late 1970s	4–5 hours/bundle	—
India		
Chamoli (hills) 1982	4–5 hours/day	3–5 km
Gujerat (plains) 1980		
(a) Forested	Every 4 days	—
(b) Depleted	Every 2 days	4–5 km
(c) Severely depleted	4–5 hours/day	—
Madhya Pradesh 1980	1–2 times/week	5 km
Kumaon Hills 1982	3 days/week	5–7 km
Karnataka	1 hours/day	3 km
Gashwal (hills)	5 hours/day	10 km
Africa		
Sahel 1977	3 hours/day	10 km
1981	3–4 hours/day	
Niger 1977	4 hours/day	

Source: Agarwal (1986)

The places these fuels are collected from are as important an issue as who collects them. It has been noted earlier that they are collected from the local environment. What does this mean? Generally speaking, where there are people there are no forests, and vice versa. In other words, whilst over 90 per cent of woody biomass in the Third World is in forest areas, 90 per cent of biomass fuels come from the agricultural landscape. It is *trees outside the forest* which are the vital source of fuelwood in the Third World.

The agricultural landscape in many parts of the Third World contains many trees, as well as shrubs and other potential fuel sources. These include both trees on farms and trees in communal areas such as steep hill slopes, river banks, grazing lands and so on (plate 5.2). Trees are an integral part of the agricultural landscape. Exactly where they are found depends on the environment and the type of agriculture in the area. In figure 5.2 a rough categorization of agricultural landscapes in the Third World is set out. Clearly the potential fuel supply of different areas will vary enormously, as will its location within the landscape.

The system of fuelwood production and consumption is set out as a flow diagram in figure 5.3. This diagram shows how the fuelwood supply of an area is part of the total woody biomass resource, with its availability limited by competing uses and access constraints (we shall come to these shortly).

Plate 5.2 Rural landscapes:
(i) livestock grazing in open savanna woodland in Kenya
(ii) trees on farms in the rice area of Sri Lanka

The woody biomass resource itself reflects the area's land-use pattern, which in turn is created by the operation of the integrated production system. This integrated production system is a product of the way in which people interact with and harness the area's environmental potential and existing resources. In other words it reflects the fusion of the environment and the land management system.

Following the diagram down the right-hand side, we can see that demand for fuelwood reflects the population's energy needs and the availability of alternative sources of fuel to meet those needs. The system represented in figure 5.3 is consequently a complex one in which people balance up their needs and priorities with the availability of resources (biomass and financial) to meet these needs. Where biomass resources are plentiful, or where cash income is sufficient and commercial alternatives available, rural people do not face energy problems. Problems arise where poor people cannot meet their varied needs for woody biomass from the supplies available to them. As such, it is not a universal problem. In some places there are no real difficulties concerning energy supplies. In others, such difficulties are widespread, affecting most or all sections of the local community. These two situations represent the extremes of a spectrum. Most areas fall somewhere in the middle, with some, but not all, people facing varying degress of difficulty in meeting their energy needs.

These problems are often hidden, in that they do not show up when one compares fuelwood demand with the local resource base. Such comparisons tell us little, as what matters is not the existence of wood and other potential biomass fuels. The key is their availability as fuels. This availability may be limited by the two factors referred to above: competing uses and access constraints.

Competing uses refer to the many other purposes, set out in figure 5.4, for which wood and agricultural residues are valued by rural people in the Third World. Many of these uses are equally vital, and the demands they place upon biomass resources severely limit their availability as a fuel (plate 5.3). Where a household is faced with shortages of biomass to meet these varied demands, something has to give. What it is will depend, firstly, on the priorities of the household and, secondly, on those whose responsibilities different tasks are. Whilst providing fuel is mainly women's work, men usually control biomass resources and ensure that their needs, for construction or whatever, are met first. Similarly, if the men can increase their cash income by selling wood, then the women's needs for fuel for household maintenance will often take second place. It is not always the fuel use which loses out; probably more typical (at least in the early stages) is the erosion of the sustainability of the woody biomass resources – quite simply, more is taken out than the environment can produce.

The quantity of woody biomass available to a household to meet its varied

1 Low population density/high woody-biomass areas	
LANDSCAPES	Tropical forests/woodlands
LOCATION	Humid tropical islands, Zaire Basin, Amazonia, Indonesia, Papua New Guinea, isolated pockets – e.g. N.E. India, Mozambique, C. America, E., S.E. & W. Africa
DOMINANT PRODUCTION SYSTEMS	Peasant agriculture, shifting cultivation, hunter/gatherer, based on communal land tenure. Large scale plantations
FUELWOOD SITUATION	Few problems
2 Low Population density/low woody-biomass areas	
LANDSCAPES	Arid/semi-arid regions High mountain regions
LOCATION	Central Asia, Sahara, Sahel Belt, much of East and Southern Africa, Central and N.W. India, Himalayan Plateau, Andean Plateau
DOMINANT PRODUCTION SYSTEMS	Pastoralism (including nomadic), isolated peasant farming
FUELWOOD SITUATION	Problems widespread, especially around permanent settlements, in areas of environmental pressure
3 Medium/high population density/high woody-biomass areas	
LANDSCAPES	High potential areas such as humid highlands, islands and coastal plains. Typically a mosaic of land uses.
LOCATION	S.E. Asia (e.g. Philippines, Thailand, Malaya), East African Highlands, Central America, Caribbean, Southern India and Sri Lanka, West Africa and pockets elsewhere
DOMINANT PRODUCTION SYSTEMS	Intensive arable production, mixture of peasant and commercial farming. Trees an important part of production system

3 (*continued*)	
FUELWOOD PROBLEMS	Frequent, especially in areas of uneven land distribution and where population densities are very high
4 Medium/high population density/medium/low woody-biomass areas	
LANDSCAPES	Varied, including river plains and deltas, transitional zones between semi-arid and high potential zones and areas around major cities
LOCATION	Indo-Gangetic Plain, S.E. Asia, N.E. and S.E. Brazil, Central America, Africa (e.g. S.E. Botswana, S. Zimbabwe, Central Tanzania, Nigeria, S. Malawi), much of Java and the Philippines
DOMINANT PRODUCTION SYSTEMS	Intensive arable production, dominated by peasant farming
FUELWOOD PROBLEMS	Common throughout these areas, acute in low productivity and in high population density areas

Figure 5.2 Agricultural landscapes and fuelwood in the Third World

needs depends on its access to the local resources. In a given place potential fuelwood supplies are not simply whatever wood is growing there, for a number of factors constrain the accessibility of part or all of these resources. These constraints on access are many and varied but can conveniently be grouped into three categories.

The first factor which influences the availability of fuelwood is the distance between the resource and the point of use. The distance which women (for it is generally women) can walk with a bundle of fuelwood is limited to, at the most, a few kilometres – perhaps up to ten in extreme circumstances. In consequence, any wood resources which lie at a distance greater than this from the users cannot be considered as part of the potential fuelwood supplies of that community. The time taken for, and ease of, fuelwood gathering is also influenced by topographical features such as hills, rivers and so on. An unclimbable slope or a deep river will act as a barrier to wood resources, whatever the physical distance involved, as it will increase

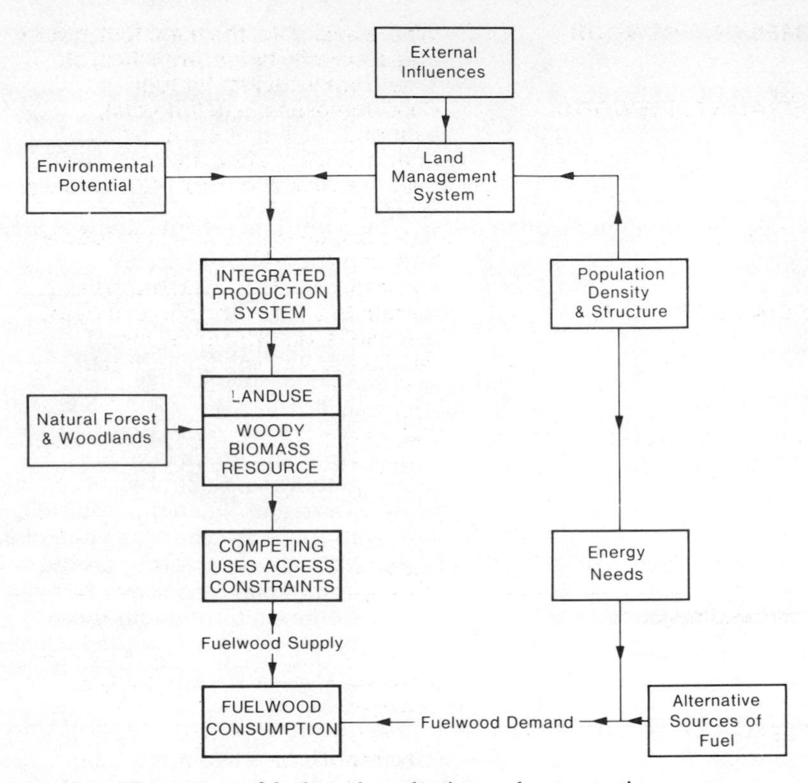

Figure 5.3 The system of fuelwood production and consumption

Tree products

CONSTRUCTION TIMBER	Wood for houses, fences, sheds, granaries, etc. The wood needs to be of good quality, with certain species preferred, tree trunks and main branches used
TOOLS AND IMPLEMENTS	Wood for agricultural implements, household utensils, boats, carts, etc., good-quality wood, with certain species preferred
FRUIT AND FOOD	Fruits, nuts, spices, edible leaves, etc. are often an important part of traditional diets, providing essential nutrients not available from staple crops

DOMESTIC FIREWOOD	Where available, the main fuel in rural areas. Usually twigs, branches, etc. Trees rarely cut just for fuel
SPECIALIST FIREWOOD	Ceremonial use (e.g. funerals), economic activities such as fish drying, brick making, tobacco curing, beer brewing, bakeries, etc. Generally larger pieces used
FODDER	Trees can be a vital source of fodder, both in pastoral areas and for domestic livestock in arable areas
SPECIALIST PRODUCTS	Certain trees are the source of dyes, medicines, natural fibres, insect repellents, etc., and often certain species have important religious significance
SHADE AND ENVIRONMENTAL PROTECTION	Trees around homesteads, in fields, along field boundaries, etc., provide shade for people, crops and animals and help protect the soil from sun, rain and erosive run-off. They are also important in denoting property rights
Agricultural residues	
CONSTRUCTION	Straw, reeds and stalks are important for roofing, making walls, fences, granaries, etc.
IMPLEMENTS AND UTENSILS	Baskets, tools, hats, agricultural implements, rope and many other essential items are often made out of crop residues
FODDER	Crop residues are often the main source of fodder for livestock, as well as providing bedding, floor covering, etc.
FUEL	Crop and animal residues are universally used as kindling or fuel for quick cooking. They are often important seasonally, and in fuelwood problem areas provide a safety-net source of fuel
SOIL FERTILITY	Manure and crop residues are a major source of fertilizer which provides essential nutrients, protects soil structure and does not have to be purchased from outside the local economy

Figure 5.4 Tree products and agricultural residues: the diversity of local needs

Plate 5.3 Multiple uses of wood: fencing and construction in Zimbabwe

significantly the collection time and consequently reduce the distances from which fuel will be gathered.

Second, a series of access constraints are derived from the pattern of land-tenure relations within an area. Within this context, the dominant issue for most rural communities is the control of land, as access to biomass resources does not depend on their physical existence but is a function of their ownership. Land which is owned and farmed by the local community is subject to private property rights. This land is frequently the main source of biomass fuels, from trees within the farms, and access to those fuels depends upon ownership of the land. Households with limited or no land will face severe restrictions upon access to these fuels. In many cases (and in particular where there is a marked uneven distribution of land) land-poor or landless families have traditional rights of access to fuel from the land of larger landowners, but such rights are eroded where biomass resources become commercialized or are under greater pressure.

Access to biomass resources within large-scale commercial farms and forest plantations or reserves is frequently highly restricted or prohibited altogether. This category of land frequently contains a large proportion of the biomass resources of many areas but, if the populace is denied access to them, they cannot be considered part of the potential fuelwood supply of such areas.

The position regarding access to the biomass resources of common land is perhaps the most complex of all. Whenever such resources are available it is normal to find a range of traditional customs and practices which regulate access to them. These are associated with different resource-management systems and are discussed below.

Third, the biomass-resource-management system in an area will limit access to the biomass resource. The management system in an area will reflect the prevailing social structure, local resource management and harvesting techniques and technologies, the range of alternative, non-fuel uses to which different species of tree are put, and customary rights and obligations concerning the use of and access to biomass resources on communal and, in some cases, private land.

Taken together, the alternative uses of wood and crop residues and limitations on accessibility will mean that the potential fuel resources available to certain households may be severely limited, even where there appears to be a plentiful supply in the local area. The key issue is who controls the local biomass resources.

The existence of widespread rural energy problems reflects this and produces a range of responses as people are forced to adapt their energy economy as supplies of fuelwood become scarcer. These responses vary according to local circumstances and are often indirect (see case study E).

Case study E

Responses to fuelwood problems

1 More careful management of the fire to use less fuel. This takes more time but can result in significant reductions in fuel consumption.
2 Travelling further and taking longer to gather fuel. As we have seen, this can have a major impact on the women's labour budgets.
3 Cooking patterns are adapted. Fewer meals are cooked (for instance, two a day instead of three), quick-cooking foods used more, communal cooking becomes more widespread and so on. In some cases these changes can have an adverse impact on nutritional levels or result in higher expenditure on food.
4 Non-fuel uses of wood or agricultural residue are cut back on where they are a lower priority than fuel use.
5 One of the most widespread responses is the extraction of materials beyond the capacity of the local environment. This reflects changed management practices, with immediate needs taking precedence over

Case study E (*continued*)

long-term sustainability. People are usually aware of these impacts but have no other choice.

6 Where wood supplies deteriorate, people switch 'downwards' to crop or animal residues. These are inferior fuels but are often more readily available. In some areas, such as Bangladesh, Lesotho and northern India, they are now the dominant fuel. As such, residues act as a fuel safety net into which people fall when wood becomes scarce.

7 Alternatively people may switch 'upwards' to commercial fuels such as kerosene and LPG. This can occur through choice, where people can afford it, but may be the only alternative, even where people are very poor. The impact on the household budget can be severe.

8 Woodfuel may become commoditized – it may change from being a free good to something people sell and have to buy. Where this happens scarcity will be acute. The people who benefit are the men and the well-off who control land and wood resources. The losers are the poor and the women, who now have to pay for what was previously available free of charge. This commodification occurs most frequently where urban fuelwood demand comes into conflict with local needs.

9 A positive response is where people begin to grow more wood. This is rarely just for fuel – the trees will provide for a range of needs – but fuel availability will certainly increase. This will occur only where land is available and the effort involved is outweighed by the benefits. One problem is the time delay – even fast-growing species such as eucalyptus take a number of years before they can be harvested. Another is the issue of control – women feel the problems but men control the land.

10 Where pressures on biomass fuel resources become acute, different aspects of the structure of roles and responsibilities in the local production system begin to break down. This can take many forms. A common one is that men begin to assist in fuel provision, either by helping to collect wood or by providing alternative resources (plate E.1). Landless households' traditional rights of access to fuel on the land of local landowners become eroded, leaving the poorest with the least access. Traditional management practices of communal lands, which are based on ecological sustainability, similarly break down where pressures become severe. These and other responses lead to the erosion of the social fabric of local communities, further threatening the viability of local production systems.

Case study E (*continued*)

Plate E.1 Fuelwood responses: a man gathers wood by cycle from distant woods in Mauritius

Some responses, such as improved fire management, are benign and sustainable, but most have a detrimental impact on some aspect of people's lives. For example, the overexploitation of local biomass resources can be a major factor in the environmental deterioration which threatens many parts of the Third World. As fuelwood problems become more acute, the responses become more drastic until, in places, they may seriously jeopardize the viability of the local production system. The list of responses to fuelwood problems in case study E starts with ones which usually occur first and progresses through to ones which are generally (but not invariably) found where the initial responses are not enough. They consequently describe a process of progressive change to the local pattern of energy provision.

The long list of responses to fuelwood problems in case study E indicates the complexity of fuelwood issues in rural area of the Third World. The problems that people face in meeting their energy needs force a range of responses which themselves contain the seeds of a worsening future or precipitate other problems. Few of these responses are sustainable. As the situation deteriorates throughout the Third World *something* must be done. Some possible responses are considered in the final chapter.

Energy and the urban poor

One of the most important features of the contemporary Third World is rapid urban growth. Towns and cities are growing at an unprecedented rate, and the urbanization process is transforming many features of the economies and societies of many regions. The growth of the urban population is paralleled by the growth of urban energy consumption. If present trends continue (which is likely) urban energy consumption will equal or surpass rural consumption within the next 20 years throughout the Third World. In regions such as Latin America, which are already substantially urbanized, this is already the case.

The forces driving urbanization are many and complex (see the volume by Drakakis-Smith in this series). Briefly, they reflect the integration of Third World countries into a global economy, as cities are the channel through which this link occurs and contain large numbers of poor people. Fuelwood is frequently the main fuel for much of the urban poor. This is particularly true in poorer Third World countries. Patterns of domestic energy consumption in urban areas are far more complex and dynamic than those of rural areas. First, for individual households, multiple fuel use is common. It is not unusual to find households using two, three or more fuels for the same purpose. This is particularly true for cooking, which dominates domestic energy consumption and is the main use of fuelwood.

Second, the structure of household energy use is different for different types of household. The main determinant here is economic status, with different income groups being characterized by different access to fuels and devices. Household size and location within the city are also important here, complicating the simple income-group relationship one would expect to find. For example, in a group of Indian cities in 1984 wood provided over 50 per cent of the energy needs for cooking and heating of the poor but less than 10 per cent of the needs of higher income groups, who relied on kerosene (33 per cent) and liquid petroleum gas (LPG) (40 per cent). The poor also used some kerosene (24 per cent of their needs), but virtually none of the more expensive LPG, which provided only 1 per cent of their needs. The same pattern was found in Kenya by the Beijer Institute, where wood was used by 82 per cent of low income households but only 6 per cent of the highest income group, who relied heavily on LPG (55 per cent) and electricity (68 per cent). Intermediate fuels were charcoal, which was used by 84 per cent of all families, and kerosene, which was used by over 50 per cent of all families except the very poor (who could not afford it) and the rich (who did not want it).

Third, urban domestic energy consumption is very dynamic, changing over time as fuel supplies, urban incomes and fuel prices change. This dynamic pattern is called an *energy transition* and can occur very rapidly if relative fuel supplies or prices change suddenly. In most cases, the transition is from biomass fuels to commercial fuels, but it can be the other way; a 'downward' transition into biomass fuels. This took place in both Sri Lanka and Indonesia in the early 1980s after the price of kerosene rose dramatically, following the second oil shock and the removal of government subsidies on kerosene.

The nature of urban fuelwood is very different from that of rural areas. In the city fuelwood is a commodity, a commercial good which is produced elsewhere, imported to the city and sold on the urban market (plate 5.4). For urban consumers fuelwood is in direct competition with other domestic fuels such as kerosene, LPG and electricity. One of the key factors determining its use is consequently its cost relative to the cost of commercial alternatives. It is very difficult to calculate these relative costs, as information on prices, efficiencies of consumption and so on is very poor. Two major studies on South Asia (by Gerald Leach of the International Institute for Environment and Development) and Southern Africa (by the

Plate 5.4 Urban charcoal sellers in Lilongwe, Malawi

ETC Foundation) show that there is no clear trend in these relative prices. In some cities firewood is considerably cheaper than commercial alternatives. In others it is somewhat more expensive. The overall picture is one of fuelwood prices often being comparable to those of kerosene, LPG or electricity.

Despite this, fuelwood is still widely used by the urban poor, even though fuelwood (but not charcoal) is generally considered a less preferable fuel. A number of factors can explain this. Not all fuels are direct substitutes for each other: some are more flexible, more convenient or preferred for different forms of cooking (slow cooking, quick frying, baking and so on). Cost is not just the price of fuel, but also that of appliances, such as stoves, which may be substantial and which must generally be paid in a lump sum (a capital cost). Access to sufficient cash to pay for expensive appliances is a real problem for many of the urban poor. Fuel use depends not just on their price, but also on their availability: the issue of *fuel security*. The way in which fuels must be paid for is important. Electricity necessitates the ability to save for monthly or quarterly bills, LPG to pay for a bottle of gas all at once, and so on. Many of the urban poor do not have the disposable income to accumulate these lump sums even if their fuel costs would be cheaper in the long run. These and other factors mean that the *cost* of different fuels consists of more than just the *price* of the fuels.

Fuelwood in urban areas, as a commodity, differs in nature from other fuels. In particular, it is produced in the petty-commodity sector by numerous producers from a range of different types of sources. In contrast, commercial fuels are frequently imported and are produced and distributed by the capitalist sector or the state. This means that the cost and scale of production of fuelwood are far more flexible than those of commercial fuels, and that in particular many petty producers will continue to extract, transport and sell fuelwood whatever the price – they frequently have little alternative.

The points discussed above form the context within which urban fuelwood must be considered. In particular, fuelwood use by the urban poor (the arena of both most fuelwood use and most fuelwood problems) can be understood only as a facet of their total pattern of fuel use within the context of the way in which they live in and cope with city life. The move to the city opens up more scope for the poor than exists in their rural homes, but it is far from Utopia for them. Urban life is harsh and, for those outside the dominant sector, imposes a series of constraints and opens only limited opportunities. The poor cope with these constraints through survival mechanisms based on their adaptability and operated through their community which maximize their limited scope and scarce income.

Fuelwood problems in Third World cities reflect the urbanization process, characterized by widespread poverty and rapid growth, outlined above.

Current energy problems fall into three broad categories: the cost of fuelwood to urban consumers, security of fuel supplies and the impact of the urban fuelwood market on the rural areas which supply that market.

The point at which cost to consumers becomes a problem is difficult to estimate. For the very poor any expenditure is undesirable, and as such any initiative which lowers their energy costs is beneficial. These costs become a real problem, however, when they reach a level, as a proportion of income, which jeopardizes their ability to provide for other basic necessities of life. Just how widespread this problem is is hard to estimate, but the data available suggest that energy costs in general and fuelwood costs in particular are rising in many Third World cities.

Fuel security is closely related to fuel costs but is important enough to be considered a distinct problem. In many countries supplies of many fuels, and in particular of petroleum-based fuels, are erratic and difficult to obtain. Much the same is true of different types of appliances. As such, the availability of fuel supplies is as much a concern to many people as their cost. This partly explains why multiple fuel use is so common, and why many people continue to use fuelwood even where commercial alternatives are, in theory, cheaper.

Urban fuelwood demand has a substantial, and frequently detrimental, impact upon the rural areas which supply the urban market. The export of wood to the city leads to competition with local needs for woody biomass, can result in the commodification of fuelwood in supply areas, and frequently creates substantial environmental damage as a result of over-exploitation and rapacious extraction techniques. All of these factors disrupt the local woody biomass production system. They may be mitigated, to an extent, by the income generated, but the producer areas typically receive only a fraction of the price the final consumers pay.

Thus the energy problems of the urban poor are different from those of rural areas. In cities the situation relates to fuels which are commodities produced elsewhere, whereas in rural areas most energy comes from non-commercial fuels which are gathered from the local environment. Just as the problems differ in nature, so do potential planning solutions. The last two chapters have focused on the nature of energy consumption in the Third World and the problems which these patterns of resource use create. In the final chapter we look at potential answers to these problems.

Key ideas

1 Fuelwood is the main fuel of the world's poor. Fuelwood problems are specific to people in places.
2 In rural areas, energy can only be understood as part of an 'integrated local production system'.

3 Biomass resources have many uses besides providing fuel. Access to these resources is restricted by physical and social constraints.
4 Fuelwood problems express themselves indirectly.
5 In cities energy problems reflect the cost and scarcity of a range of fuels. They are different to rural energy problems.

6
Third World energy: options for the future

The energy problems facing the Third World are daunting. As we have seen, they retard development and threaten the survival of many of the world's poor. Such problems present formidable challenges to governments, international agencies and individuals who are concerned with and responsible for energy planning. There are no easy answers to these problems, as their root causes are deeply embedded in the overall pattern of development and underdevelopment of the contemporary Third World.

Total solutions are difficult to identify, but in most cases *something* can be done. This chapter considers the main policy options open to energy planners in the Third World. In particular, it examines the types of policy which seem to be appropriate to particular energy problems.

Before doing this, it will be useful to review quickly the main energy problems facing the contemporary Third World. One set of problems relates to the process of development at a national level. The cost and availability of supplies of commercial fuels retard this development process. The cost of fuel (especially oil) imports, the high levels of capital investment needed to set up energy generation and distribution systems, problems with the reliability of fuel supplies and the sheer cost of energy all push up production costs for industry and the cost of living for consumers. They also limit the foreign exchange available for other vital imports and the capital available for other necessary investments. Energy expenditure by consumers lowers demand, and consequently economic growth, in other sectors of the economy, further retarding development prospects.

The second set of energy problems revolves around the ability of the poor to provide for their basic needs. Fuel costs and shortages are part of the crisis of survival which they face. These are largely, but not exclusively, a question of the decrease in the availability of non-commercial biomass fuels. For the poor, the fuelwood crisis leads to additional strains on their labour budgets, the commercialization of previously free resources, the deterioration of their local environmental resources, and a range of other, indirect impacts which disrupt their pattern of production and, ultimately, jeopardize their survival prospects.

These are the problems. What are the solutions? In a nutshell, if one is confronted with an energy problem there are three options available:

1 Conservation: by using fuel more efficiently the same amount of work can be done with less fuel (or more work done with the same amount). The efficiency of use of fuel largely depends on the machinery in which it is used. It is consequently both a technological and an economic question, as technology defines efficiency and capital availability defines the ability to afford the technology.
2 Substitution: if a specific fuel or range of fuels is too expensive, unavailable or has to be imported, in certain cases an alternative fuel supply can be substituted. This depends upon the existence of cheaper and available alternatives which can be used to power the same machinery or alternative machinery to do the same job.
3 Supply enhancement: if existing supply sources are insufficient and/or too expensive, then in some cases alternative supplies of the same fuel can be developed. This depends on the existence of suitable resources for development at a price which is affordable.

These three options, conservation, substitution and supply enhancement, should be borne in mind as we look at each of the main sets of energy problems in the Third World and consider both what has been done in the past and what should be done in the future.

Energy and development

The basic problem in this context is the cost and availability of commercial fuels, with imported oil being of particular importance. Continued development necessitates increasing supplies of these fuels. Of crucial importance are the industrial and transport sectors, with household use also growing in significance as urbanization and living standards grow.

Conservation

The pattern of use of most fuels severely limits conservation possibilities. Pricing policies can and do limit use, but the adverse impact of such policies disproportionately hits less well-off households and can also jeopardize the viability of businesses, which are already struggling in an unfavourable international climate, by raising their costs. The other route to conservation is by introducing more efficient technologies. The great problem here is technological dependency. Almost without exception, transport equipment (cars, trucks, etc.), household goods and much industrial plant are designed and built in the First World, to serve the needs and conditions of the First World. The Third World has to use expensive, imported machines just as they have to use expensive, imported fuels. Often industrial plant which is redundant in the First World is shipped to and used in the Third World. Invariably machines are kept running for as long as they can, and the older they get the less efficient they are. So conservation is limited by the lack of availability of efficient, appropriate machinery. Although more energy-efficient technology has been developed in the First World (this has been particularly true since 1974), Third World countries cannot afford to scrap existing machines and import a new technology; the capital costs are beyond their means.

Substitution

Substitution possibilities are more widespread and have received considerable attention in recent years. As far as commercial fuels are concerned, the main possibilities appear to be replacing imported fossil fuels with indigenous fossil fuels (we consider this below, under supply enhancement), nuclear power and new and renewable sources of energy such as solar power, wind power and novel forms of biomass fuels.

Although still confined to a handful of countries (see figure 6.1, table 6.1), the use of nuclear power in the Third World has grown substantially in recent years. Nuclear power's advocates would argue that it provides a cheap and reliable source of electricity which does not require either local energy resources or major fuel imports. The impact of the oil shocks in the mid and late 1970s made nuclear power appear far more attractive to a number of oil-import-dependent Third World countries. This was particularly true for what are known as the 'Newly Industrializing Countries' (NICS), as these Asian and Latin American states were pushing for substantial economic growth based on industrialization, and hence on rapid increases in the availability of energy supplies.

One of the major constraints on the use of nuclear power is the size of the electricity system. In general, no more than 10 per cent of grid capacity should come from any one generating plant. Given a standard nuclear reactor size of about 600 MWE, this restricts nuclear power to countries with electricity systems of 6,000 MWE or over; a size which only 30 or 40 Third World countries will achieve by the year 2000. Efforts by reactor manufacturers such as Westinghouse to produce viable smaller reactors have so far proved unsuccessful.

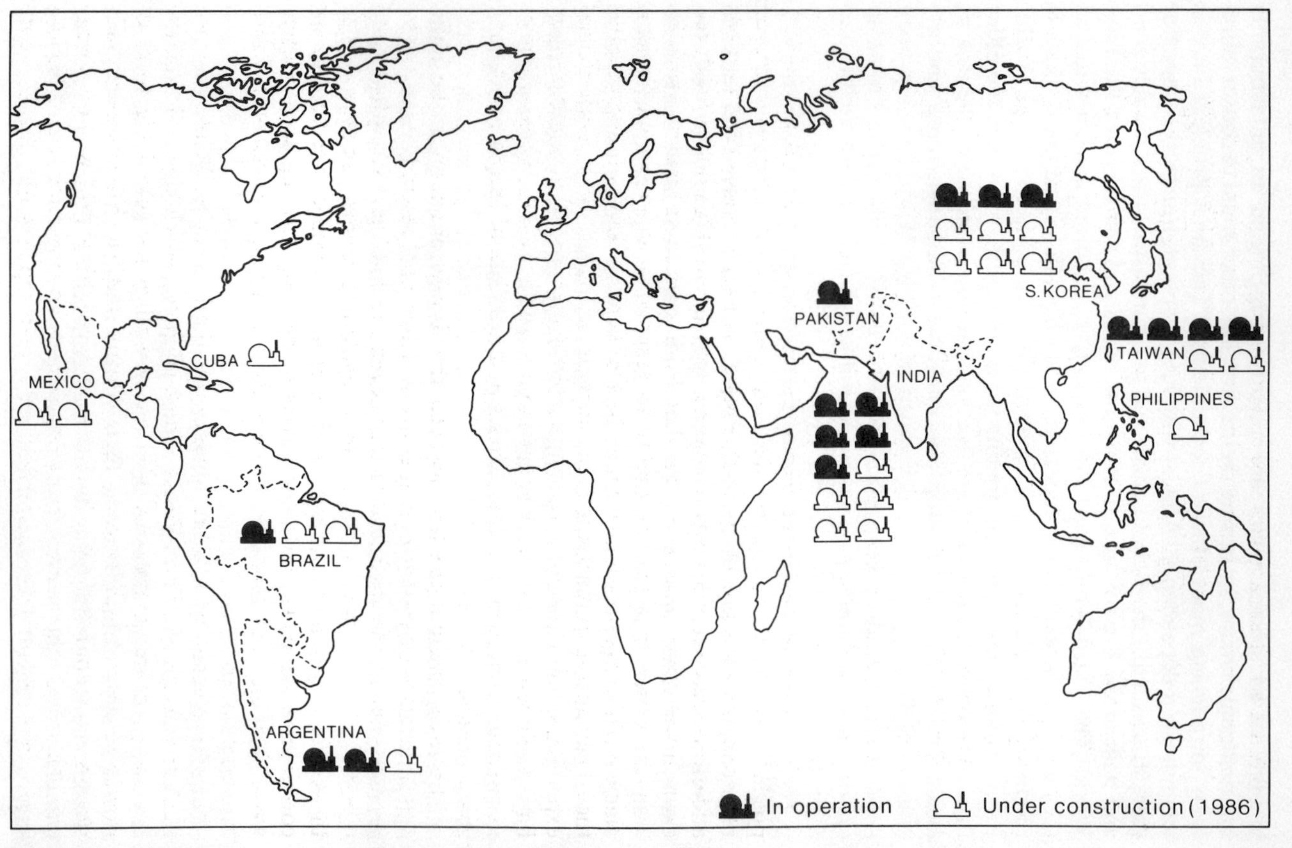

Figure 6.1 Nuclear reactors built and under construction in the Third World
Source: *Guardian* 16 May 1986

Table 6.1 Nuclear power in the Third World

	In operation		*Under construction (1986)*		*Total MWE*
	Number	*MWE*	*Number*	*MWE*	
Argentina	2	935	1	692	1,627
Brazil	1	626	2	2,490	3,116
Cuba			1	408	408
India	5	1,030	5	1,100	2,130
Mexico			2	1,308	1,308
Pakistan	1	125			125
Philippines			1	621	621
South Korea	3	1,789	6	5,474	7,263
Taiwan	4	3,110	2	1,814	4,924

Source: *Guardian*, 16 May 1986

As is true in the developed world, the economic arguments for nuclear power in the Third World are far from convincing. The massive capital requirements are particularly problematic, as they increase debt burdens and divert scarce capital away from other investments. The power produced tends to be more expensive than that from alternative sources, and the resulting system has little flexibility to respond to changing fuel prices. Nuclear Third World countries may be less dependent on oil imports, but they become totally dependent on imports of nuclear fuels and external expertise, spare parts and so on. With a few exceptions, such as India which has developed her own nuclear technology and expertise, nuclear power necessitates compliance to the policies and attitudes of supplier firms and governments.

The environmental problems associated with nuclear energy in the Third World are also a matter of great concern. Chernobyl reminds us all of the potential impact of a nuclear disaster in even a technologically sophisticated society such as the Soviet Union. In a developing country, safety systems, the ability to respond swiftly to a disaster and economic pressures to cut corners all make it even more hazardous a technology than in the developed world. The disposal of nuclear waste is also a major, and as yet unresolved, area of concern.

In simple resource and economic terms there is no justifiable case for nuclear power in the Third World. Those countries which have followed the nuclear path find themselves with an expensive and unreliable form of power which is largely beyond their control and which has substantially increased debt burdens. Why have they followed this path? Two non-energy reasons provide the answer: its prestige (many governments regard nuclear power as a symbol of their modernity and sophistication) and, more ominously, its military applications. It is no accident that nuclear power has been adopted mainly by military or authoritarian regimes (and has been rejected when civilian rule is restored, as in Brazil, Argentina and the Philippines). Countries such as India, Pakistan, Israel and South Africa are known to have developed bomb capabilities, although they often deny this. The same is suspected of Taiwan, South Korea and others. The threat of nuclear weapons proliferation is the most important reason for (and argument against) the spread of a form of energy provision which cannot be justified on economic or resource terms.

New and renewable sources of energy have received a great deal of attention over the past 20 years or so, with many studies emphasizing both their apparent abundance and their appropriateness to the energy needs of the Third World. Innumerable machines have been designed to harness their perceived potential, and numerous programmes to disseminate such appropriate technology have taken place. Despite a sustained effort over many years, the impact of new and renewable sources of energy has been marginal and fraught with technical difficulties. This has produced an inevitable reaction; the 'new and renewables' are becoming increasingly rejected as unsuitable to the needs of the Third World.

The 'new and renewables' include a wide range of types of potential energy resources. Some, such as wave power, tidal power and ocean thermal energy conversion (OTEC) are of such limited extent and fanciful design that they need not be taken seriously here. Others, such as solar, wind, hydro, novel uses of biomass and to an extent geothermal, must be treated more seriously, as their distribution is more widespread and the technology associated with their use more established. Despite this, even these energy sources are limited in their impact. None has proved to be a widespread substitute for either fossil fuels or fuelwood.

Part of the reason for their failure to make a significant impact is that their development has been almost wholly led by technologists, who have mostly seen solutions to Third World energy problems in terms of a 'technical fix'. The machines produced have certainly been ingenious and novel but have been far from 'appropriate' in social or economic terms. The following quote from a conference on Solar Energy for Developing Countries held, significantly, in London in 1986 sums this vision up well:

> The new technology, which converts sunlight directly to electricity, photovoltaics, is technically and economically appropriate to achieve the required transformation in the quality of (village) life.

A brief examination of the main new and renewables will reveal why their potential as substitutes for existing fuels is so limited.

The potential of *solar power* seems abundant. The Third World lies mostly in the tropics, with long periods of intense sunlight. By its nature, however, solar power is dispersed. Capturing the resource is the problem. This has advantages where the needs are similarly dispersed but limits its potential for concentrated, large-scale users such as industry or power generation in urban areas. The second problem is the type of energy produced. Solar power is ideal for space or water heating (using flat plate collectors), but the Third World rarely needs space heating, and hot water is a luxury rather than a necessity. Solar water heaters have proved effective for specific uses such as heating water for tourists in hotels in countries such as Turkey, Sri Lanka and Mauritius. Elsewhere they are of little relevance.

Photovoltaic cells convert sunlight directly into electricity. Their potential would appear more widespread, but to date the cost of the cells has proved prohibitive. They reflect a classic technology trap. Cost will remain high until they are mass produced. Mass production is unlikely without a mass market in the First World, but the lack of both need and resource potential in these dominant countries preclude such a market developing. Catch 22. Photovoltaics do have specific uses, even at current prices. This is particularly true for remote locations (mountainous areas, deserts, islands, etc.), where they can provide essential power for health clinics, radio stations and so on (plate 6.1). Again, this is a specific application, not a generalizable substitution for current energy demand.

Plate 6.1 Solar energy for development: photovoltaic cells used to power a remote radio station in Mali

Hydro-power is the most widely used and proven form of new and renewable energy. Large-scale HEP projects are found extensively throughout the Third World and often contribute significantly to national energy demand. Such schemes create their own problems (see case study F), but their potential is nevertheless largely realized.

Case study F

The Mahaweli HEP Project in Sri Lanka

One of the main sources of electricity in a number of Third World countries is large-scale hydro-electric power (HEP) projects. Some of these large dam schemes are huge; the Aswan High Dam in Egypt is a famous example, the Itaipu Dam on the Parana River in Brazil is the world's largest (with a generating capacity of 12,600 MWE), and other massive schemes include the Volta Dam in Ghana, Zimbabwe's Kariba Dam and the Cabora Bassa complex in Mozambique.

Typically, big dams are combined power generation/irrigation projects which aim to provide electricity for cities and industry and control water for agricultural development. Their scale is huge, involving massive commitments of capital, taking years to complete and affecting the landscape of thousands of square miles of countryside. The reservoirs behind such dams are often hundreds of kilometres long and involve submerging tens of thousands of hectares of land.

A classic example of the big dam approach was opened in 1985 in Sri Lanka. The Accelerated Mahaweli Development Programme was first started in 1970 but was pushed forward when the government of Julius Jayarwadene gained power in 1977. The scheme involves three main dams along the Mahaweli river, which flows out of the central highlands area in Sri Lanka to the sea on the east coast near the port of Trincomalee (see figure F.1). The cost of the project, originally estimated at £700 million, has inflated to well over £2 billion, the bulk of which Sri Lanka has provided herself. The Mahaweli complex will double Sri Lanka's electrical generating capacity from 560 MWE to 1070 MWE and, on completion, is planned to provide controlled irrigation to 210,000 hecatres, much of it virgin land in the eastern dry zone which is being developed under a massive settlement programme. A number of countries have provided about £400 million in aid to the project but, as is often the case, this is a double-edged sword. For example, Britain has provided about £100 million loans and grants but British companies have received over £150 million in contracts on the project.

The Mahaweli scheme has become a major political issue in Sri Lanka

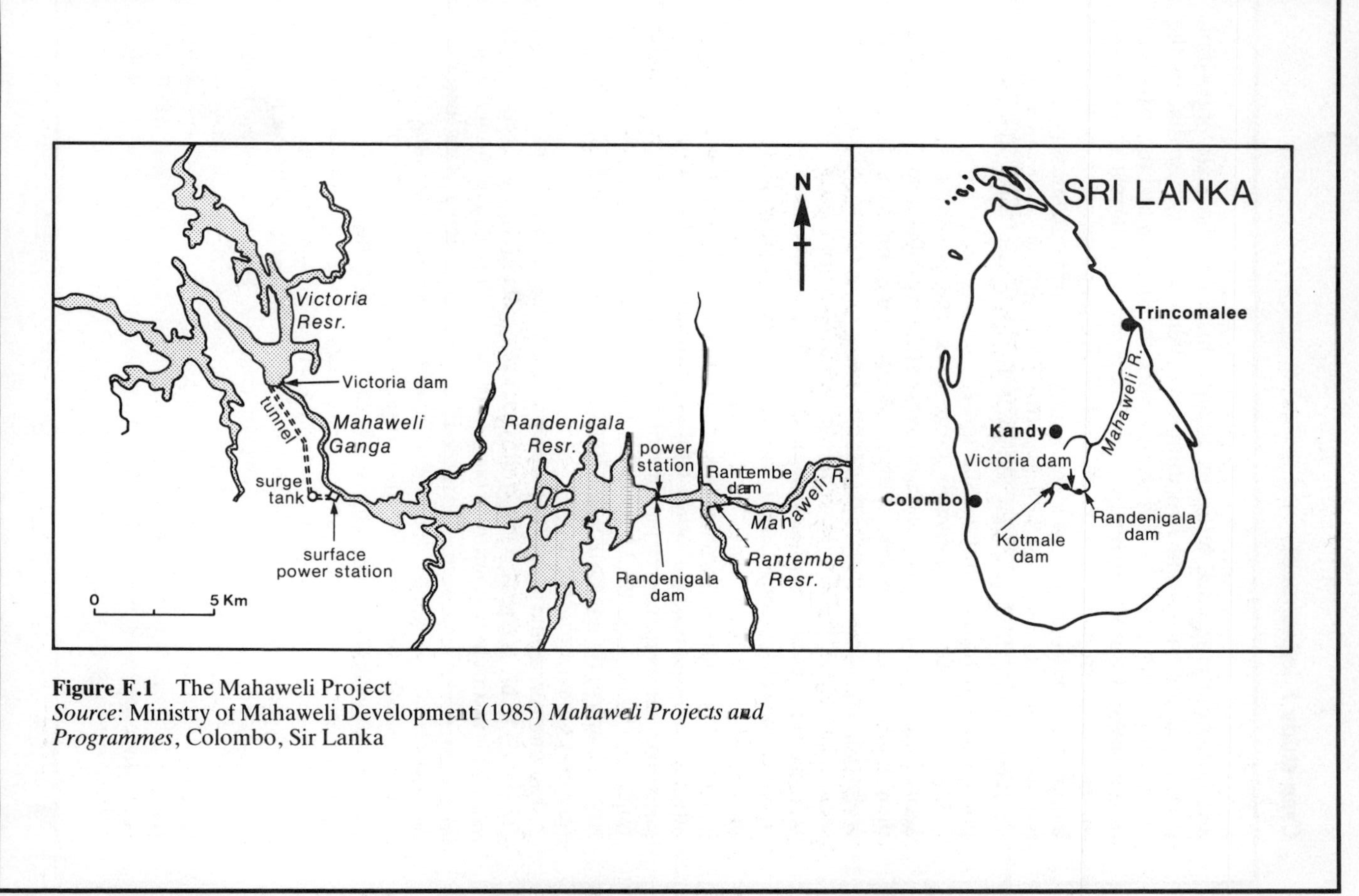

Figure F.1 The Mahaweli Project
Source: Ministry of Mahaweli Development (1985) *Mahaweli Projects and Programmes*, Colombo, Sir Lanka

Case study F (*continued*)

and is fervently opposed by many groups, including the main opposition party. The electricity and irrigation have been seen as essential to the current economic policy, which stresses industrialization and the expansion of agricultural output to remove food imports, provide cheap food and increase cash crop exports. Prior to 1977 frequent power cuts hindered industry and discouraged outside investment. The land-settlement programme has mainly involved moving Sinhalese from the south-west into an area split amongst Tamils, Sinhalese and Muslims. In strife-torn Sri Lanka this has proved to be a highly controversial issue.

The effects of the Mahaweli project on the government and local agriculture have been similarly controversial. Some 5,000 families have lost their land to the reservoirs. Offers of compensation have been widely criticized. The same is true of the resettlement programme, which is widely seen as being politically, rather than economically, motivated. The claimed increases in agricultural production are unlikely to be realized and already many of the settlement schemes are suffering economic and environmental problems.

The concentration of Sri Lanka's capital resources in this project has had adverse consequences for other vital capital needs within the country. In particular, alternate energy and infrastructure projects have been starved of funds as the cost of the Mahaweli scheme has inflated. The capital costs have significantly increased Sri Lanka's debt burden and drained her foreign-exchange reserves.

The environmental consequences of the scheme are the biggest cause of concern – a characteristic Mahaweli shares with all big dam projects. Large-scale deforestation has occurred both in the watershed and in the settlement areas (in which fuelwood provision is a major problem). This has been accompanied by rapidly increased erosion, producing serious siltation problems in the reservoirs for the future. Many of the settlement areas are already suffering salination of the soil because of poor irrigation management, again a problem found in many dam schemes. A range of wildlife habitats has been seriously disrupted, including breeding grounds for migratory birds and river fish. The effects of fertilizers and pesticides in the irrigated areas on water quality is already causing concern. The canal system is easily clogged by weeds and provides an ideal habitat for many disease-bearing insects. Malaria, once wiped out in Sri Lanka, is proving a major health hazard. The disruption of the water cycle is likely to have long-term adverse consequences, with the water table dropping in some areas and rising in others.

Case study F (*continued*)

All of these economic and environmental problems have been found in other parts of the Third World where large dams have been developed. Poor economic returns, increased health hazards and environmental deterioration are the counterpoints to increased power generation and irrigation capacity. To governments, power for the cities and irrigation for cash crops appear to be more important than the health of people or the environment. The prestige and political credibility associated with large projects is important too. Big dams provide large quantities of cheap electricity. Whether the non-economic costs are worth paying is a matter of individual judgement.

A form of hydro-power which is novel, and which has received much attention, is mini-hydro, essentially small-scale HEP schemes in rural areas to serve local needs. Such schemes originated, on a substantial scale, in China and received favourable reports there. Their use elsewhere has been more limited. The advantages of mini-hydro is its dispersed nature, its negligible environmental impact and, above all, the low level of investment it requires. However, disadvantages include the amount of power generated, which is often not enough to make the effort worthwhile, the technical requirements, which necessitate expertise usually unavailable at village level, the high cost of the power generated and the seasonality of flows of many rivers which interrupt the power generation. Mini-hydro plants have been installed in some countries outside China (mostly in mountainous regions, for example in Nepal, Northern India and Colombia). To date they have not been successful, and prospects for their future development are not good.

Wind power has similarly failed to fulfil expectations. The wider exploitation of this resource is limited by similar constraints to those discussed above. The end-uses for wind power are confined to activities such as pumping water; useful, but hardly a substitute for oil use in industry or transport. Similarly, the cost of wind turbines, the high level of technical skills needed to maintain them, the erratic availability of the resource and other factors all limit its use to specific niches in the energy economy of the Third World.

Biogas is a methane-based fuel produced from the anaerobic digestion of organic material (generally dung) by micro-organisms. Again developed extensively in China, where communal digesters have proved successful, this technology has been tried in many countries. There are numerous smaller, household digesters in India and trial projects throughout the Third World. The cost of the digesters, the effort involved in collecting dung and the cost and complexity of gas-using appliances have limited their applicability. In parts of India where well-off landowners have built digesters, their greatly increased needs for dung have resulted in poor landless families losing their rights to collect this fuel for traditional use, a classic example of where an 'appropriate' technology has helped the rich at the expense of the poor.

The production of *alcohol-based substitutes* for petrol or diesel from plants such as sugar cane, cassava and sunflowers has attracted much attention ever since Brazil began to implement an oil-substitution policy in 1975. Other countries such as Zimbabwe have plans to copy the Brazilian experiment, and ethanol could provide an effective way of using cash crops whose export value has declined. By 1985 over 2 million vehicles in Brazil were running on alcohol alone, and the remaining 8 million used a 20 per cent alcohol–petrol blend. In 1984 over 90 per cent of all cars sold were powered by alcohol, and the programme has given a boost to Brazil's car industry. The fuel is expensive, however, costing the equivalent of over $40 per barrel. In the early 1980s this was competitive with oil, but declining oil prices since 1986 have cast a cloud over the alcohol programme. Overall alcohol for fuel is certainly an option for the future but it is probably not economic today.

Dendrothermal power involves using wood to produce electricity. The idea is that biomass-rich areas can use spare resources to generate the power they need. The problem is that biomass-rich regions contain few people and consequently do not need the power. The largest experiment in dendrothermal power, in the Philippines, has proved disastrous. The wood was to be grown by small farmers who were given loans and incentives to provide the fuel for a series of small-scale power plants. The problem has been that, if the price paid to the farmers has been high enough to reward adequately their efforts, then the cost of the fuel has been too high to make the electricity competitive. The capital costs have also proved prohibitive. The project is only kept going by government subsidies, and this experience has discouraged other governments from repeating the experiment.

Overall the story of new and renewable sources of energy in the Third World is one of disappointment. These resources appear to have great potential as fuels, and are widely used for specific applications but they do not appear to offer a viable alternative to the dominant fuels in the Third World. This disappointment is misplaced. Such resources never should have been considered for this role: the dream of a solar economy is just that, a dream. We have seen that these resources have much to offer in a series of specific circumstances. Their greatest potential is as a means of generating electricity which can be cheaper and more readily available than fossil fuel-based generation in specific circumstances. Apart from large-scale HEP schemes, these circumstances occur in dispersed or remote locations where specific needs exist and where the national power system does not reach. Other examples of specific roles for new and renewables can be identified. Case study G, which looks at energy in small islands, demonstrates these points well. This narrower but important developmental role for new and renewables is regarded as heresy by many people who have staked their careers on their development, but the sooner their real potential is realized, the sooner this potential can become a reality.

Case study G

Energy planning for small island nations

Of the 140 or so developing countries in the world, a significant proportion are islands that have a population of less than five million and a commensurately small gross national product. Small island nations have been characterized as having open and vulnerable economies. As Selwyn (1978) has said, they are 'small, poor and remote'. Limited domestic demand, narrow export bases, high levels of dependence, extremely restricted access to capital, extreme lack of infrastructure and skilled labour and, in many cases, limited and fragile environments all combine to produce extreme development problems in these island states. A central aspect of these problems is energy.

Apart from a few exceptions, such as Trinidad and Tobago which are fortunate enough to possess economically exploitable deposits of fossil fuels, island developing countries depend on two principal sources of energy – indigenous biomass resources and imported petroleum products. Table G.1 gives percentages for the energy sources of a number of oil-importing island countries in the Caribbean, Indian and Pacific oceans. In most cases both oil and biomass are major contributors to the country's energy supply, with each serving distinct sectors of the economy.

The limited land mass of islands imposes obvious contraints on the expansion of supplies of biomass fuels, restrictions that population growth makes more acute. Similarly expansion of imported commercial energy supplies is restricted by the burden this would place on the balance of payments. In addition, due to the small volumes delivered, island communities often pay a higher price for their oil products than larger customers.

The use and sustainable production of biomass fuels is obviously an issue of major importance for small island countries which is rarely given the

Case study G (*continued*)

Table G.1 Energy sources for selected small islands

	Area (km²)	*Population (000)*	*Annual per capita consumption (kgoe)*	*Primary energy supplies (%)*			*Oil imports as a % of export earnings*
				Oil	*Biomass*	*Others*	
St Lucia	616	124	320	61	39	—	17
St Vincent and the Grenadines	388	113	220	53	45	2	9
Haiti	27,756	5,000	270	17	79	4	29
Mauritius	2,000	983	380	38	60	2	16
Seychelles	400	65	350	91	9	—	29
Solomon Islands	28,000	246	430	34	66	—	22
Fiji	18,371	658	370	43	55	2	10
Vanuatu	15,000	126	NA	22	78	—	40
Tonga	671	101	NA	30	70	—	66
Western Samoa	2,840	159	NA	39	60	1	94
Kiribati	790	60	NA	49	51	—	150

Source: Kristoferson, O'Keefe and Soussan (1985)

attention it deserves, and which may one day prove to be a greater barrier to development than the very serious one posed by the cost of sustaining and expanding oil imports. It is clear that small island developing countries are experiencing major constraints on development because of energy-related problems. All need to enhance available energy supplies, and to do so wherever possible through the development of indigenous energy resources.

For biomas fuels, strategies such as improved forest management, fuelwood plantations of rapidly growing species for dendrothermal (wood-fuelled) electricity production and conservation policies, such as agroforestry and improved charcoal production, clearly have great potential. Nevertheless they are always limited by the obstacle of competing demands in a confined geographic space.

A number of island states may have the potential to develop small deposits of oil or gas to provide for their needs. The problem is their lack of resources to explore and actually develop these deposits. The multinational oil companies are not interested unless there is a chance of large deposits for export. Without access to international capital, expertise and technology

Case study G (*continued*)

island states have little, if any, chance to develop fossil fuels for their own needs.

So what of the alternative new and renewable sources of energy – solar, wind, hydro, wave, geothermal and ocean thermal energy conversion (OTEC)? For many island countries the theoretical potential of such energy resources is great. Most lie in the sunny tropics and many have strong and reliable winds and waves, hilly terrains and reliable rainfall (and, hence, good small-scale hydro potential), the majority are in ocean areas with large vertical temperature differences (for OTEC) and are frequently of volcanic origin (which indicates geothermal potential). As yet these renewable resources are virtually untapped. This is partly because the required technologies are either unproven or not yet competitive with existing commercial alternatives. In addition, there is a paucity of research and development focused on the specific needs and potentials of small, remote and poor countries. The scale of production, capital requirements and infrastructural needs of the technologies that have been developed are often beyond the needs and means of individual small island states.

Small island states provide an example of places where specific needs and problems could be met by a novel and innovative approach. In these states alternative energy supplies are entirely appropriate to provide for the needs of small, remote and resource-poor communities. These opportunities can only be met, however, if the international community is willing to provide the capital and expertise; on their own, individual island states are just too small, poor and remote to cope.

Supply enhancement

For many Third World countries, the most realistic prospect for meeting their commercial energy needs comes from the development of indigenous fossil fuel deposits. A number of Third World countries are, of course, already substantial oil exporters. The declining price of oil has led to countries such as Mexico, Indonesia and Nigeria increasingly valuing the non-oil sectors of their economies and being willing to divert oil exports to fuel internal development. Such an approach entails costs, but these are worth paying. Other countries (for example, India and Brazil) have actively developed their oil production since 1974 and are increasingly able to meet at least part of their needs from domestic production.

Coal and natural gas are indigenous energy resources of considerable and increasing importance for many Third World countries. *Coal* deposits have been developed in countries as diverse as China, India, Zimbabwe, Colombia and South Korea, and in these countries make a major contribution to national development. Coal is particularly important for electricity generation, industry and, in some cases, transport. As such, it is an ideal development fuel. It can also be used in households where fuel-wood scarcity is a problem. A further group of countries (for example, Botswana and Indonesia) have considerable but as yet poorly developed coal reserves. For these countries coal is undoubtedly a fuel for the future. The main problem with its development is the extremely high capital costs involved. Problems of access to technology and expertise and severe environmental impacts are also important issues. Despite such problems, if coal can be developed, it is usually highly competitive and offers considerable scope for replacing fuel imports in sectors such as industry and power generation.

Production of *natural gas* in the Third World is growing more rapidly than that of any other commercial fuel; output rose by 11.2 per cent per year between 1980 and 1985 and is expected to continue to grow by at least 10 per cent through the 1990s. The World Bank listed 47 Third World countries with potential for natural gas production in 1980. Of these a number, including Colombia, Bolivia, Algeria, Bangladesh, Pakistan and Thailand, are already significant producers and in others natural-gas production is planned or under development. The role of gas can be expected to increase significantly in the future.

Natural gas is used mainly in electricity generation and in industry. It is ideally suited for these sectors, providing a ready substitute for oil imports, and it can also be used in urban households. Countries such as Bangladesh and Thailand have managed to cater for growing energy demand without increasing oil imports. This has had beneficial effects on their economies – in Bangladesh's case holding off a bankruptcy which seemed certain in the early 1980s.

Where it is possible then, production of oil, coal and natural gas by Third World countries is the most immediate and effective way out of their commercial energy problems. These possibilities obviously depend on resources being available. One of the great problems Third World countries face is the exploration for, and development of, their resources. The oil and gas exploration process is totally controlled by First World multinationals which are interested only in the discovery and development of large, commercially viable deposits for the world market. They are also highly selective as to where they will look, being unwilling to work in countries which are considered to be politically risky.

Only about 5 per cent of the land area of the Third World has been properly explored for fossil fuels. The potential of the rest is unknown. A similar problem for Third World countries is that many of their fuel deposits are small – too small to be considered worth developing by the giant oil companies who are interested only in major fields to supply the world market.

Such fuel sources, those in unexplored regions or small deposits, could make a major contribution to the commercial energy problems Third World countries face. Discovering and developing them are beyond their control, however. There have been calls for an international effort to assist in developing this potential, but so far little has been done. First World governments and international agencies such as the United Nations and World Bank must lead such efforts, either by providing the capital and expertise needed or by putting pressure on the oil companies to change their policies on exploration and control of technology. Until this happens, the potential of indigenous fossil-fuel deposits, which are the best bet for helping the Third World out of the development trap commercial energy produces, will not be realized.

Energy for the poor

Planning for the energy needs of the Third World's poor must recognize the local context of their energy use. As such, it will necessarily differ between urban and rural areas. Towns are points of fuel consumption but not production. This is particularly true for woodfuel, which in urban areas is commoditized and in direct commercial competition with other fuels. In contrast, in rural areas most fuels are non-commercial, local goods. Rural localities are sites of both production and consumption where energy problems must be placed in the context of the local production system.

Planning for rural energy

Rural energy planning in the Third World is essentially concerned with sustaining production and consumption of biomass fuels. Biomass fuels are locally produced but are rarely specifically for fuel use. These fuels are usually either a residue from other productive activities or gathered from natural woodland areas where resources are exploited, but no direct production occurs.

The locality-specific characteristics of rural production systems form the context within which energy planning must take place. Any planning options which ignore or seek to replace this reality will fail. Such failures litter the history of rural energy planning in the Third World. Inappropriate technical fixes, misguided forestry programmes and ineffective conservation strategies are the norm. The exceptions, of which there are few, are successful energy-planning exercises. We can develop these points using the three categories, conservation, substitution and enhanced supply, outlined above.

The scope for *conservation* strategies in rural areas is extremely limited. Where fuelwood becomes scarce people become conservation conscious and save fuel by better fire management. This apart, conservation strategies centre on the development of improved cooking stoves (plate 6.2) which save fuel by burning more efficiently. With an open fire typically only about 10 per cent of the energy is effectively used. Many designs of stove have achieved 30 per cent or greater levels of efficiency in the laboratory.

Plate 6.2 An improved cooking stove in Sri Lanka

In practice, however, the efficiencies achieved in ideal laboratory conditions are rarely transferred to the field. Most programmes pay insufficient attention to the mass production and dissemination of their stoves. In most cases, improved stoves have to be paid for and can be expensive. Expecting the rural poor to spend scarce cash on a stove to replace the free open fire is unrealistic. Many stove designs are inappropriate to people's needs, as they do not take account of the way people cook and roles of fires as sources of light and foci of social activity. Fires also help to keep down insects. These factors have combined to undermine the effectiveness of stove programmes. Their future role is being assessed by many planning agencies, but the impact of conservation on rural fuelwood problems will be, at best, marginal.

Much the same is true for *substitution policies*. Replacing fuelwood with alternative fuel invariably means a switch from a locally produced free good to a commercial good imported from outside the local community. In practice, these commercial fuels are likely to be oil products such as kerosene and LPG (the problems with new and renewable sources of energy have already been discussed).

If the rural population can afford the fuels, this 'energy transition' will occur anyway except where fuel distribution systems are so bad that people are discouraged by erratic supplies. In such cases, the role of the planner is to improve fuel-distribution systems to provide the fuel security people need. This means that fuel switching in rural areas is a possibility only where the level of economic prosperity is such that people are willing and able to start paying for their fuel needs. By definition, this excludes the poor and disadvantaged, who face the worst energy problems in rural areas.

If conservation and substitution are discounted, it leaves *increased biomass fuel supplies* as the main response to rural energy problems. In other words, energy planning must increase the total biomass resources, improve the proportion of existing biomass resources available as fuel or both. Mixed land-use and multiple uses of trees within the total production system is the model with which energy planners must operate. In other words, rural energy planning must seek to strengthen and build upon the local production system described above.

Increasing biomass fuel supplies has traditionally been equated with planting trees. This can be promoted under government, communal or individual management and occur on state, communal or private land. Projects based on communal or state management which are aimed at increasing fuelwood supplies are usually called either 'social forestry' or 'community forestry'. The intention of such projects is to establish woodlots – essentially miniature plantations – on plots of communal or state land in agricultural areas. In most cases the planning model has been top-down, with government officials doing the planning and communal participation, such as it is, usually confined to the government experts informing people of their plans. Even where local people are involved, it is typically the economically and politically dominant men who control proceedings whilst those who need the wood most, women and the landless, are largely excluded.

This type of rural energy planning has rarely been successful. Agarwal (1986) describes a number of such failures, including projects in Ethiopia where labourers planted trees upside down, in Niger where saplings on communal grazing land were uprooted to allow cattle to graze, in Tamil Nadu (India) where eucalyptus saplings were uprooted and in the Philippines, where Tinggian tribal communites deliberately started fires in areas taken over by the forestry department for community forestry. However, active resistance is unusual; more commonly people tend to ignore such projects, which consequently fail because of the lack of expected participation by the local community. For example, the Bangladesh Community Forestry Project was set up in the early 1980s to provide fuelwood in north-west Bangladesh, which has one of the world's most acute rural energy crises. Woodlots were to be established in village communal lands, along road and canal sides and on government-owned land. Millions of seedlings were planted, but no attention was paid to their care and survival rates of the seedlings were only about 5 per cent.

The basic problem with this sort of project is that it does not meet people's needs. Land is developed for one use only, to grow fuelwood trees. Often this runs counter to people's needs; for example, communal grazing land is alienated from people, or tree species which grow slowly, but which are valued for a range of uses, are replaced by fast-growing but single-use trees which people do not know and do not wish to know. In other words, they fail because those in charge do not recognize the complexity and diversity of the way rural people interact with their local environment to provide for a range of needs.

Projects which have involved individuals planting trees on private land have often been more successful. This approach is called 'farm forestry' or 'agroforestry' (see case study H). The basic idea is to recognize and strengthen the role of trees as a component of Third World farming systems. This can be based on existing practices, for example, providing technical and material support to encourage the role of trees as field barriers, around homesteads and so on (plate 6.3). It can also aim to introduce new agricultural practices whose aim is to strengthen and diversify the farming system. Agroforestry is the growing of trees and crops together on one piece of land. Agroforestry is not new; indigenous agroforestry characterizes many traditional farming systems in the Third World.

The aim of agroforestry projects is either to introduce new crop–tree combinations or to extend this type of farming to areas in which it is not traditional. If the right combinations of trees and ground crops are used, a farmer will find his total output increases significantly. Woodfuel is not necessarily the sole, or even the main, output – wood for sale, fruit crops and so on can be the principal product and extra fuelwood a by-product. If the circumstances are right (see case study H), the potential of this approach is great, but it must be built from the existing production system and agroforestry is by no means the miracle answer some suggest. It will work best in high potential areas, requires viable markets for the commercial

Plate 6.3 Farmers cultivating trees in Zimbabwe

component of the crop combinations and may necessitate levels of investment beyond the means of poor farmers. There are many agroforestry schemes being developed throughout the Third World. It is as yet too soon to see whether they will work, but the approach certainly offers hope.

Case study H

Agroforestry

Agroforestry schemes – the use of farmland for the combined production of tree and ground crops – have been identified as an appropriate way to increase fuel supplies without taking up scarce land or seriously disrupting existing patterns of production. The key to the successful development of agroforestry is the correct selection of crop–tree combinations, but it is also essential that the range of products created reflects people's needs and, crucially, that the demands of the crops and the trees reflect both the local environmental potential and the availability of labour in the household. For example, the Oxfam Agroforestry Project in Burkina Faso has proved successful because it was based on detailed consultations with local farmers, addressed a range of problems including environmental degradation and

Case study H (*continued*)

crop production and relied on techniques which local people could readily adopt, which used local materials and which called on labour in the slack dry season rather than in the wet season when none was available.

Leguminous species of tree such as *leucaena* are nitrogen fixers – they increase soil fertility in suitable conditions. These and other species can be 'nutrient pumps', absorbing nutrients from lower levels in the soil and recycling them through leaf decay. Some ground crops benefit from the shade provided by trees. For example, in Kenya, coffee which is grown amongst shade trees requires fewer fertilizers and is less demanding on the soil. Similarly, in the oases of southern Tunisia, a range of fruits and vegetables can be grown under the date palms which are the main commercial crop (plate H.1). The leafy palms protect them from the fierce desert sun and cut greatly the amount of irrigation needed. This can be particularly important during the early stages in the growing cycle. Where the crop–tree combinations are right the farmer will find his *total* output greatly enhanced. In the Dominican Republic, for example, most land

Plate H.1 Indigenous agroforestry: date palms and ground crops in Kebili, Tunisia.

Case study H (*continued*)

holdings are less than one hectare in size. By carefully combining a range of tree and ground crops the small farmers are able to achieve a better use of the soil, water and light resources to grow crops for food and for sale as well as provide fuel, fodder, medicines, herbs and wood for building and other needs.

As far as woodfuel production is concerned, agroforestry has a number of advantages. First, woodfuel demand is already supplied from trees on farmland. Agroforestry strengthens existing patterns of production and is readily recognized by farmers as meeting their needs. A good example of this is the USAID's agroforestry outreach project in Haiti. The project was aimed at easing Haiti's acute fuelwood problems by encouraging small farmers to increase wood production on their farms. The approach adopted was flexible, permitting farmers to decide when, where and how to plant seedlings provided by the project. Voluntary agencies were used to reach the farmers, and in four years over 27 million seedlings were distributed to 110,000 farmers.

Second, present fuelwood gathering techniques concentrate on wood that is small in diameter. This offers the possibility of developing short rotation 'energy trees' which could easily be integrated into existing production systems. Third, agroforestry can provide a continuous supply of wood, which can be 'harvested' much like any other crop by techniques such as coppicing and pollarding. Fourth, the trees do not have to be exclusively fuelwood trees but can provide fruit, poles for construction, environmental protection and other outputs. This makes them far more acceptable to the male farmers, who control land but are not necessarily worried about fuelwood shortages. Finally, as we have seen, agroforestry can increase the land's total productivity.

The Kenya Woodfuel Development Programme, outlined in case study D, is a good example of where this approach has successfully reached the people. World Bank projects in Gujerat and Uttar Pradesh (both in India) have similarly been very successful in encouraging farmers to increase tree production. Recent studies have shown that, in many countries (for example, Malawi, Tanzania, Indonesia, Pakistan), farmers are responding to wood shortages by expanding tree production without the encouragement of external planners. People are responding to problems in the best way they can. In these and other instances producing extra fuelwood is not the sole reason for increasing tree planting. In many cases the commercial value of wood poles is the main incentive, in others tree fruits are commercially valued. Extra woodfuel is a by-product but is produced nevertheless.

Case study H (*continued*)

Agroforestry consequently has many advantages. It is, however, by no means a universal panacea. First, the environmental potential of an area must be suitable for the intensification of production through increased tree growth. In particular, many of the favoured fast-growing species of tree, such as eucalyptus, are extremely demanding on soil and water resources. They will fail if tried in places where these demands cannot be met.

Second, increased production of wood may not solve the energy problems of the poor. If the people who have energy problems have no access to land or land resources, then growing more trees may not help them. For example, in the Gujerat World Bank project quoted above, wood production increased greatly, but it was all on the land of large farmers who sold the wood for construction poles or wood pulp. The income they received was substantial, encouraging them to look for wood elsewhere. This led to decreased wood availability for the poor. As such, another World Bank project succeeded in benefiting the well-off at the expense of the poor.

Third, the commercial component of the crop combinations must be marketable, and the output must be sufficient to meet the farm's commercial needs. Similarly the farmer must have access to some capital and be able to support the disruption of production while the new combinations are established.

Fourth, where land and/or labour pressures are such that all aspects of the production system are under pressure, then growing more wood may mean producing less of other vital products – such as food. In such circumstances something has to give. Whether it is fuel or another basic need will depend on local circumstances, but in such areas there is no easy answer to the development crisis the poor face.

Agroforestry has great potential. This is most likely to be realized where it is developed on the basis of people's knowledge and needs. What this needs is a genuine bottom-up approach, in which the people themselves make decisions – not external 'experts' who usually understand neither the local population's needs nor their relationship to the local environment.

A final approach to increasing biomass fuel supplies is to impove the management of natural woodlands, an approach appropriate to many semi-arid or mountainous areas, where population densities are low and the environmental potential is limited. In such areas pastoralism, rather than arable agriculture, is generally the dominant production system. The potential for encouraging private tree growth is limited by both the environment and the system of land ownership and management. In such areas, private land tenure is rare and nomadic movements across tracts of land are common. Communally owned woodlands or bush savannas are the source of grazing for animals, wood for fuel and many other products which serve people's needs.

Pastoral production systems break down where land is alienated from people for settled agriculture (often cash crop plantations, usually with the best land being lost), and where increasing needs for cash (to buy vital products, pay taxes, pay for schooling, etc.) lead to increasing numbers of livestock which in turn intensifies pressure upon the land. Under such circumstances traditional management techniques of the rangelands break down or are increasingly inappropriate. The role of rural energy planners in such areas is to improve these management techniques to ensure a sustainable yield of fuelwood from the rangelands, and this in turn means ensuring environmental stability, and in particular stopping the spread of desertification which follows from the loss of tree cover and the overgrazing of semi-arid rangelands.

The way in which this can be done will depend on local circumstances, but techniques include improving the rotation of grazing patterns, excluding livestock from specified areas and allowing natural regeneration, introducing better wood extraction techniques, replanting environmentally crucial areas, such as gullies and watersheds, or opening up wider areas of rangeland to nomadic pastoralists. All of these responses require the complete and enthusiastic involvement of the pastoralists themselves. Again, a top-down planning approach is useless. No one knows the potential and limitations of the local landscape better than the local people themselves. To work, planning must build from this local knowledge and, again, seek to strengthen rather than replace the local production system.

Energy for the urban poor

The rapid growth of cities throughout the Third World means that urban energy problems will continue to increase rapidly. Woodfuels (including charcoal) are the main fuels of the poor, but they are a commercial good. Their acquisition is a problem because of overall cost (including the cost of energy-using devices) in relation to that of other basic necessities of life. Many households use several fuels because supplies of individual fuels are erratic. Fuel security is also an important issue to the urban poor. Even where the poor do not face serious energy problems now, rapid urban growth means that the future must be planned for. Urban energy planning must have a future dimension. The final aspect of urban energy is the impact of the urban market for fuelwood on rural areas. This commercialization of rural fuelwood leads to environmental destruction and often threatens the needs of the rural population in the supply areas.

The main approaches to urban energy problems have been either to subsidize alternative fuel supplies such as kerosene or electricity, or to increase fuelwood supplies by establishing peri-urban plantations. Both these approaches are expensive, and both have largely proved ineffective.

Subsidies on fuel prices cost too much for poor Third World governments. They lead to low fuel-use efficiencies, and inevitably bring more benefit to bigger users (the rich, businesses). They amount to using a blunderbuss where a telescopic sight is needed. To work, urban (and rural) energy policies must be closely targeted to specific communities in need.

Peri-urban plantations have been much favoured by organizations such as the World Bank and FAO. They have universally failed and are now being abandoned as an approach. The basic problem is that the wood they produce is far more expensive than that already on the urban market. Peri-urban plantations require land close to the city and are expensive in labour, capital and foreign-exchange terms. They can rarely compete with fuelwood from natural woodlands, where the production costs are not paid.

The need for a new approach is now recognized. Increased fuelwood supplies for the city could come from the improved management of existing supply sources and the opening up of alternative sources. Transport costs are a major problem, but with careful management to ensure environmental sustainability the needs of the city need not be in conflict with rural needs or the protection of the environment.

Where urban fuelwood problems seriously affect the urban poor, it is because scarcity has led to high prices. In these cases commercial fuels such as LPG, electricity and kerosene are as cheap or cheaper than wood or charcoal. People continue to use wood because of fuel-supply insecurity or appliance costs. Policies aimed at encouraging fuel switching to commercial fuels from wood can be cheap and effective in these circumstances. Such measures try to improve fuel-distribution systems, provide cheap and available stoves, set up rationing systems and so on, an approach which provides people with the fuels they want at a price they can afford. Of course, for the very poor, any expenditure on fuel is a burden, but it is beyond the realistic resources of the state totally to subsidize fuel for the urban poor.

Conservation, via improved stoves, has more to offer in cities than in rural areas. This is because people buy stoves anyway and, as long as the improved models are as cheap and adaptable as existing ones, will readily adopt them. Urban stove programmes have been set up in a number of countries and have often achieved noteworthy success in terms of the numbers distributed.

Overall, energy problems in cities must be approached as part of the general urban planning system. The plans must recognize the real nature of the problems the urban poor face, be realistic about the limited resources available to the planning system and target their efforts to the specific problems of defined communities. If the planning system does this, it has some chance of success. If not, urban energy planning, like so much planning in the Third World, will achieve little and, at times, make things worse.

Conclusions

Energy planning in the Third World is clearly a complex issue. The scale and nature of energy problems vary from place to place but throughout the Third World are becoming a major development issue. The solution to these problems lies in building on what is already there: seeking to harness local resources and develop sustainable production of fuel which provides for people's real needs and does not disrupt people's lives. Such goals are hard to achieve but, unless they are the targets, the dual energy crises facing the Third World will continue to undermine development prospects and jeopardize the survival prospects of the poor.

Key ideas

1. Energy policies can solve problems by fuel conservation, substitution or supply enhancement.
2. Nuclear power and new and renewable sources of energy cannot replace oil as the main fuel for development or wood as the main fuel of the poor.
3. Rural energy planning must strengthen, not replace, the local integrated production system.
4. Urban energy planning must provide for the poor, plan for the future and reduce the impact of urban fuelwood demand on rural supply areas.

Review Questions

Chapters 1 and 2

1 How can we define the quantities of a resource available both now and in the future?
2 Why does the geography of mineral production differ from the geography of mineral consumption?
3 What has happened to mineral prices over the last 25 years? Why have these changes occurred?

Chapter 3

1 What is the significance of the data in table 3.1?
2 Why do giant multinationals dominate the world's minerals industry?
3 What are the costs and benefits of mineral production for Third World countries?
4 What tactics have Third World countries adopted to improve the benefits they receive from mining? Why have they failed?

Chapter 4

1 What are the characteristics of energy resources?
2 What are the main fuels used in different sectors of the economy?
3 What is the dual energy crisis?

Chapter 5

1 Where are fuelwood problems found?
2 What will affect the availability of biomass fuels in any one of the agricultural landscapes described in figure 5.2?
3 How do energy problems of the urban poor differ from those of the rural poor?

Chapter 6

1 What is the role of new and renewable sources of energy in the Third World?
2 What problems do Third World countries face in developing their fossil-fuel deposits?
3 Why has rural energy planning failed, and how can it succeed?

References

Chapters 1–3

Blunden, J. (1985) *Mineral Resources and their Management*, London, Longman.
Caldwell, M. (1977) *The Wealth of Some Nations* London, Zed Press.
Fernie, J. and Pitkethly, A. (1985) *Resources, Environment and Policy*, London, Harper & Row.
Mikdashi, Z. (1976) *The International Politics of Natural Resources*, Ithaca, NY, Cornell University Press.
Mikesell, R. (1980) 'The structure of the world's copper industry', in S. Sideri and S. Johns (eds).
Nwoke, C. (1987) *Third World Minerals and Global Pricing*, London, Zed Press.
Odell, P. (1986) *Oil and World Power* (8th edn), Harmondsworth, Penguin.
O'Riorden T. and Turner R. (1983) *An Annotated Reader in Environmental Planning and Management*, Oxford, Pergamon.
Partizans (1985) *RTZ Uncovered*, London, Partizans.
Rees, J. (1985) *Natural Resources: Allocation, Economics and Policy*, London, Methuen.
Rio Tinto-Zinc (1985) *The Rio Tinto-Zinc Corporation PLC Annual Report and Accounts*, London, Rio Tinto-Zinc.
Sampson, A. (1975) *The Seven Sisters*, London, Hodder & Stoughton.
Sideri S. and Johns S. (eds) (1980) *Mining for Development in the Third World*, New York, Pergamon.

Tanzer, M. (1980) *The Race for Resources*, London, Heinemann.
Timberlake, L. (1987) *Only One Earth*, London, BBC/Earthscan.
World Commission on Environment and Development (1987) *Our Common Future*, Oxford, Oxford University Press.
World Resources Institute (1986) *World Resources 1986*, New York, Basic Books.

Chapters 4–6

The journal *Ambio*, vol. 14, nos 4–5, contains a series of articles on energy in the Third World. See in particular articles by: Bradley *et al.*, Kristoferson *et al.*, Soussan and O'Keefe and Chadwick *et al.*
Agarwal, B. (1986) *Cold Hearths and Barren Slopes*, London, Zed Press.
Commoner, B. (1976) *The Poverty of Power*, New York, Alfred Knopf.
Earthscan (1981) *New and Renewable Energies*, London, Earthscan.
Eckholm, E., Foley, G., Barnard, G. and Timberlake, L. (1984) *Fuelwood: the Energy Crisis That Won't Go Away*, London, Earthscan.
ETC (1987) *Wood Energy Development: a Study of the SADCC Region*, Leusden, Holland, ETC.
Kristoferson, L., O'Keefe, P. and Soussan, J. (1985) 'Energy in small island economies', *Ambio*, vol. 14, no. 4–5: 242–4.
Leach, G. (1986) *Household Energy in South Asia*, London, International Institute for Environment and Development.
Odell, P. (1986) *Oil and World Power* (8th edn), Harmondsworth, Penguin.
Ramage, J. (1983) *Energy: a Guidebook*, Oxford, Oxford University Press.
Selwyn, P. (1978) 'Small, poor and remote: islands at a geographical disadvantage', Discussion Paper 123, Brighton, IDS, Sussex University.
Soussan, J. and O'Keefe, P. (1985) 'Biomass energy problems and policies in Asia', *Environment and Planning A*, vol. 17: 1293–301.
Soussan, J., Ferf, A. and O'Keefe, P. (1984) *Fuelwood Strategies and Action Programmes in Asia*, Bangkok, AIT.
Timberlake, L. (1987) *Only One Earth*, London, BBC/Earthscan.
World Bank (1986) *World Development Report 1986*, Washington DC, World Bank.
World Commission on Environment and Development (1987) *Energy 2000*, London, Zed Press.
World Resources Institute (1986) *World Resources 1986*, New York, Basic Books.

Index